PHOSPHORIMETRY

The Application
of Phosphorescence
to the Analysis
of Organic Compounds

THE APPLICATION
OF PHOSPHORESCENCE
TO THE ANALYSIS
OF ORGANIC COMPOUNDS

PHOSPHORIMETRY

by M. ZANDER

RÜTGERSWERKE UND TEERVERWERTUNG A.G.
CASTROP-RAUXEL, GERMANY

Translated from the German by THOMAS H. GOODWIN

DEPARTMENT OF CHEMISTRY
UNIVERSITY OF GLASGOW, SCOTLAND

NEW YORK AND LONDON *1968*

ACADEMIC PRESS INC.
111 Fifth Avenue, New York, New York 10003

United Kingdom Edition published by
ACADEMIC PRESS INC. (LONDON) LTD.
Berkeley Square House, London W.1

LIBRARY OF CONGRESS CATALOG CARD NUMBER: 68-18686

PRINTED IN THE UNITED STATES OF AMERICA

PREFACE

Phosphorimetry is an interesting spectro-analytical method that, although only recently developed extensively, has already been used frequently in such varied disciplines as pharmacology, biochemistry, tar chemistry, petrochemistry, polymer chemistry, and food chemistry. Many different types of organic compounds show phosphorescence of long duration when excited with ultraviolet light under suitable conditions. As this phosphorescence has spectral properties characteristic of the particular compound, a number of analytical applications become possible.

No previous major publication has dealt exclusively with this analytical procedure. The present monograph seeks to fulfill this need. It is directed chiefly to analysts, but organic chemists, biochemists, and others whose work involves spectroscopic relationships and techniques will find much to excite their interest.

In the first part of the book an introduction to the theoretical and experimental foundations of the phosphorescence of organic compounds is given. Here the phosphorescence properties of numerous classes of compounds and of many individual substances are discussed. The second part is concerned with phosphorimetry itself: its techniques and instrumentation are discussed and many examples of its application to different fields of analysis are given. Every effort has been made to provide complete coverage to mid-1967 of the widely scattered literature on the analytical applications of phosphorescence.

The book has materialized chiefly from the practical work done by me in an analytical laboratory.

For their generous support during the development of the work on phosphorescence spectroscopy I thank the directors of Rütgerswerke und Teerverwertung A. G., Frankfurt am Main. I am obliged to Dr. T. H. Goodwin of the Chemistry Department of the University of Glasgow for many suggestions and to the staff of Academic Press for their outstanding cooperation.

Castrop-Rauxel, Germany M. ZANDER
January, 1968

Translator's note: I gratefully acknowledge the advice of various colleagues, particularly that of Dr. J. K. Tyler who read the translation.

T. H. GOODWIN

Glasgow, Scotland

CONTENTS

PART II

ANALYTICAL APPLICATIONS OF PHOSPHORESCENCE (SPECTROPHOSPHORIMETRY)

Chapter 3. Experimental Procedures

Chapter 4. Examples of the Application of Phosphorimetry

PART I / THE PHOSPHORESCENCE OF ORGANIC COMPOUNDS

CHAPTER I / THEORETICAL AND EXPERIMENTAL FOUNDATIONS

1.1. Singlet–Triplet Intercombination Transitions

In the ground states of unsaturated organic compounds the orbitals of lowest energy are each occupied by two π electrons with antiparallel spins* (Fig. 1a). The excited states arise from the promotion of an

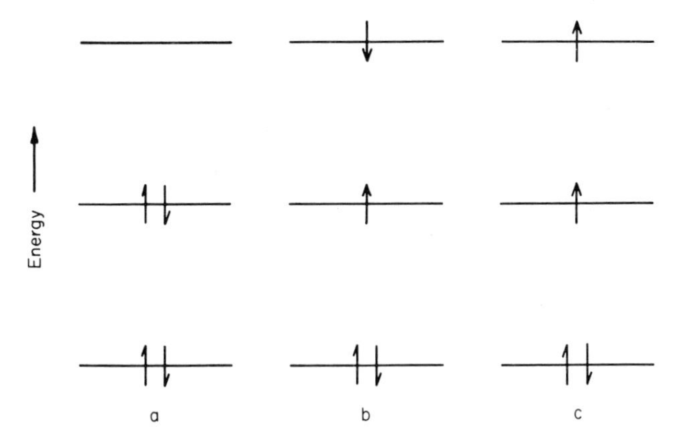

Fig. 1. Molecular orbital scheme for a singlet ground state (a), a singlet excited state (b), and a triplet excited state (c).

electron from the highest occupied orbital to one that is unoccupied. The spin angular momenta of the two electrons that are now in singly occupied orbitals are no longer restricted by the Pauli principle and

* "Free radicals" are exceptional; in this summary we are not interested in them.

their spins may therefore be either antiparallel (Fig. 1b) or parallel (Fig. 1c). The state represented by Fig. 1b is described as a singlet state and that represented by Fig. 1c as a triplet.

Both these terms are derived from classical atomic spectroscopy. It is well known that every electron in an atom has an orbital angular momentum and a spin angular momentum. In light atoms the mutual interactions of all the orbital angular momenta on the one hand, and of all the spin angular momenta on the other, are large (Russell–Saunders coupling) compared with the interaction of the orbital angular momentum and the spin angular momentum of each individual electron (*jj* coupling). The orbital angular momenta then add together to a total orbital angular momentum L and the spin angular momenta to a total spin angular momentum S. Finally S and L add together vectorially to give a total angular momentum J. If all the electrons occur in pairs with antiparallel spins, then $S = 0$ and the spins of the electrons contribute nothing to the total angular momentum. We obtain thus a single value for J, namely $J = L$, to which there corresponds a single electronic term and thus a single spectral line. Hence we describe the electronic state prevailing when $S = 0$ (antiparallel spins) as a singlet state. If, however, two electrons have parallel spins (as is the case in Fig. 1c), then we obtain a total spin angular momentum $S = 1$. Vectorial addition of L and S when $L > S$ (as, for example, with the p electrons of carbon) now leads to three different values of J according to whether S is added to or subtracted from L or remains without influence on it. To these three values of J correspond three somewhat different energies that are revealed in the spectrum as three neighboring lines. Hence we describe the electronic state prevailing when $S = 1$ (parallel spins) as a triplet state.

Actually the line separation in the triplet state is extremely small for light atoms; it increases with increasing atomic number. For the carbon compounds of interest to us it is too small to be observed in the band spectrum.* We therefore envisage the triplet state of an organic compound, in spite of its original meaning, as approximating a single state.

Let us suppose that the states represented by Figs. 1b and 1c correspond to the lowest excited states of the molecule. As the triplet state of Fig. 1c is always of slightly lower energy than the singlet state of Fig. 1b,

* The separation can, however, be confirmed by ESR measurements (see p. 14, l.c.[24]).

in which the same two orbitals are occupied (see Section 1.3), the term scheme shown in Fig. 2 is obtained. Irradiation with light of suitable wavelength causes an electronic transition from the ground state S_0 to the singlet excited state S_1. Such radiation-induced transitions between terms having the same multiplicity (singlet–singlet transitions) are of high probability. They lead to intense absorption bands. The lifetime of the excited state S_1 is very short and the emission of energy in the return from S_1 to the ground state (fluorescence) follows with great rapidity.

The situation is quite different in the transition $S_0 \rightarrow T_1$. Here we are concerned with a radiation-induced transition between two terms of different multiplicities (singlet–triplet).

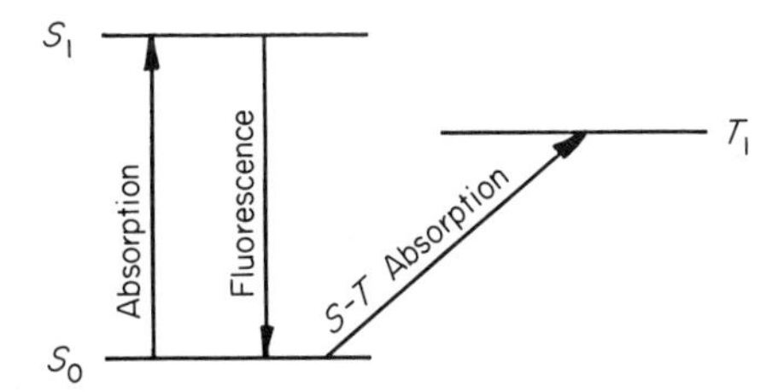

Fig. 2. Simplified term scheme.

If we assume that in our system Russell–Saunders coupling prevails exclusively and the singlet and triplet states are therefore quite unperturbed, then theory asserts that the radiation-induced transition $S_0 \rightarrow T_1$ is entirely forbidden (intercombination restriction).[1] This clearly means that the uncoupling of electron spins on excitation is extremely improbable, and is known as the spin conservation rule.

The probability of a transition induced in a molecule by radiation can be derived by quantum mechanics as an expression that, in addition to an electric dipole operator, involves the orbital and spin wave functions of the ground and the excited states.[2] This "transition moment integral" becomes zero if the electronic states defining the transition are of different multiplicities, and in particular if they are pure, i.e., unperturbed triplet and singlet states. Such unperturbed states, which are rigorously defined by reason of their multiplicities are, however, extremely rare. Generally the pure states "mix" to a certain extent—as we say intuitively—so that

[1] See W. Finkelnburg, "Einführung in die Atomphysik," Springer, Berlin, 1948.

[2] See W. Kauzmann, "Quantum Chemistry," p. 581. Academic Press, New York, 1957.

triplet states take on some singlet character and conversely singlet states some triplet character. The quantum mechanical description of this perturbation is given for a triplet state by

$$\psi_T = \psi_T{}^0 + \{A\}\,\psi_{Si} \tag{1}$$

Here $\psi_T{}^0$ is the wave function of the pure unperturbed triplet, ψ_{Si} that of a pure unperturbed singlet, ψ_T is the wave function of the perturbed triplet, and A a coefficient of mixing that indicates the extent to which the singlet state mixes with the triplet state. If we now substitute the wave function of Eq. (1) for the perturbed triplet excited state into the expression referred to previously for the transition probability, we find that although the pure triplet part of Eq. (1) still contributes nothing to the transition moment, the small singlet term in Eq. (1) now gives rise to a small value for the moment. Thus, quantum mechanically, there can occur a transfer of radiation between a singlet ground state and a perturbed triplet excited state just like that of a permitted, though of course very weak, singlet–singlet transition.

Cases in which a relaxation of the restriction on intercombination occurs have long been known in atomic spectra.[1] They are observed especially with heavy atoms. In these, *jj* coupling between the orbital and spin motions of the individual electrons occurs simultaneously with Russell–Saunders coupling and so makes the reversal of the spin easier. A familiar example is the mercury line at 2537 Å that corresponds to an intercombination transition.

This perturbation of the triplet state, which relaxes the rigorous classification of states according to their multiplicities, also applies to organic molecules in connection with the mechanism of spin–orbit coupling.[3] Thus the mixing coefficient A, which in Eq. (1) defines the extent to which the triplet state mixes with the singlet, is, on the other hand, a spin–orbit operator by which the probability of the singlet–triplet transition is determined. Spin–orbit coupling increases with increasing atomic number of the heaviest atom in the molecule and with decreasing difference between the energies of the participating singlet and triplet states. Both the parameters involved are included in the spin–orbit operator.

[3] D. S. McClure, *J. Chem. Phys.* **17**, 905 (1949); **20**, 682 (1952); E. Clementi and M. Kasha, *J. Mol. Spectry.* **2**, 297 (1958); for a comprehensive review, see M. Kasha, *Radiation Res.* Suppl. **2**, 243 (1960).

Turning back to the term scheme of Fig. 2, we can now expect that the radiation-induced transition $S_0 \rightarrow T_1$ is not entirely forbidden even in compounds that are composed only of carbon and hydrogen. It will have a small but definite transition probability. The absorption bands corresponding to the transition $S_0 \rightarrow T_1$ will have very small molar extinction coefficients, and the lifetime of the lowest triplet state will be great and energy will be emitted slowly as the molecule returns to the ground state. A further important characteristic of singlet–triplet intercombinations should be their great sensitivity to the intra- and intermolecular effects that influence the spin–orbit coupling.

The singlet–triplet absorption of benzene was observed for the first time by Sklar[4] in 1937. A very thick layer of the liquid shows a weak absorption band at about 3400 Å that Sklar correctly assigned to the transition from the ground state to the triplet state of the molecule. Lewis and Kasha[5] confirmed this observation and found values of ca. 4×10^{-4} for the molar extinction coefficients. The singlet–singlet absorption band of longest wavelength for benzene is 10^5 times as intense. Later investigations, especially those of Evans,[6] revealed that because of the presence of dissolved oxygen Sklar[4] and Lewis and Kasha[5] had not detected the unperturbed *S–T* transition and that the intensity of this is really considerably lower still.

Because of the enhancement of the spin–orbit coupling in halogen-substituted benzenes these compounds are expected to be more favorable than benzene for observing *S–T* absorption. McClure *et al.*[7] found, for example, with *p*-dibromobenzene, a long-wavelength, structured absorption band with extinction coefficients of ca. 2×10^{-1} attributable without any doubt to the $S_0 \rightarrow T_1$ transition. The same authors also noted the *S–T* absorptions of several halogen-substituted naphthalenes. Especially clear results were obtained with β-iodonaphthalene, the *S–T* absorption spectrum of which lies between 4800 and 3800 Å, with molar extinction coefficients of about 1×10^{-1}. Since then the singlet–triplet absorption spectra of numerous compounds have been measured.

The chief experimental difficulty in measuring the singlet–triplet absorption spectra is found in the very great ease with which they can

[4] A. L. Sklar, *J. Chem. Phys.* **5**, 669 (1937).
[5] G. N. Lewis and M. Kasha, *J. Am. Chem. Soc.* **67**, 994 (1945).
[6] D. F. Evans, *Nature* **178**, 534 (1956).
[7] D. S. McClure, N. W. Blake, and P. L. Hanst, *J. Chem. Phys.* **22**, 255 (1954).

be simulated or falsified by the (permitted) singlet–singlet absorption bands of impurities present in even the smallest proportions. Nevertheless, by the help of a series of criteria yet to be considered, singlet–triplet absorption transitions may be differentiated from singlet–singlet.

The discussion so far has been confined exclusively to electronic transitions that occur by taking in or giving out radiant energy. There

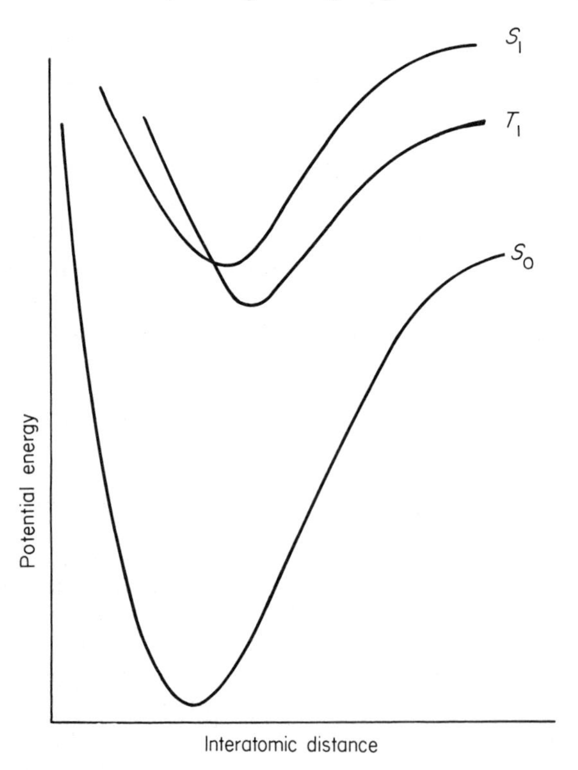

Fig. 3. Scheme of potential energy curves for ground state, and for singlet and triplet excited states.

also occur in an excited molecule *radiationless* transitions between different electronic states. The term scheme of Fig. 2 may be translated (for the hypothetical case of a diatomic molecule) into a potential energy curve, as shown in Fig. 3. As will be seen, the potential energy curves of the singlet and triplet excited states intersect in the region of higher nuclear vibrational energy; in this region a radiationless transition is possible from the singlet excited state to the triplet excited state. For the more complicated molecules in which we are interested here, the two-dimensional potential energy diagram must be replaced by a polydimensional figure having the same significance. We shall not give here the details of the theory, which we owe chiefly to Teller.[8]

Radiationless transitions proceed with great speed between excited states, but the radiationless transition from an excited state to the ground state of the molecule is, in general, much slower. This is deduced from the experimental observations and can be attributed to the fact that the definite crossing of potential energy curves* that characterizes radiationless transitions is found most frequently between the excited states, whose potential energy curves lie close together, rather than between an excited state and the ground state, since the latter lies considerably lower in the majority of molecules. In molecules in which the term difference between the excited states and the ground state is also small radiationless transitions ought to occur in greater proportion, and this is indeed the case. Franck and Sponer[10] have extended Teller's concept of radiationless transitions in complex molecules. They make use of Förster's theory of "intermolecular radiationless energy transfer."[11] The observation that radiationless transitions to the ground state are rare is then explained, because according to Förster's theory energy transfer is possible only between electronically excited states.

Radiationless transitions in a molecule can take place between electronically excited states of either the same or different multiplicity. The process of radiationless transition between two terms of the same multiplicity (singlet–singlet and triplet–triplet) is described as *internal conversion*,[12] that between two terms of different multiplicity (singlet–triplet) as *intersystem crossing*.[12] The processes are distinguishable by

* In addition to this, the most favorable case for radiationless transitions, those in which the curves, touch or approach closely without crossing, must also be considered. See l.c.[9]

[8] E. Teller, *J. Chem. Phys.* **41**, 109 (1937).
[9] See Th. Förster, "Fluoreszenz organischer Verbindungen," p. 86ff. Vandenhoeck & Ruprecht, Göttingen, 1951.
[10] J. Franck and H. Sponer, *J. Chem. Phys.* **25**, 172 (1956).
[11] Th. Förster, *Ann. Physik* [6] **2**, 55 (1948).
[12] M. Kasha, *Discussions Faraday Soc.* **9**, 14 (1950).

their speeds; for internal conversion these have an order of magnitude of 10^{13} sec^{-1} and for intersystem crossing about 10^7 sec^{-1}. From this it may be shown that both radiant and radiationless intercombinations are less probable by a factor of about 10^6 than transitions in which no change in multiplicity ensues.

Figure 4, which is developed from Fig. 2, is the term scheme of an unsaturated organic compound. Radiant transitions are indicated by full lines and nonradiant by broken lines. The electronic transition from the ground state to the singlet excited state S_2 (path 1 of Fig. 4) takes

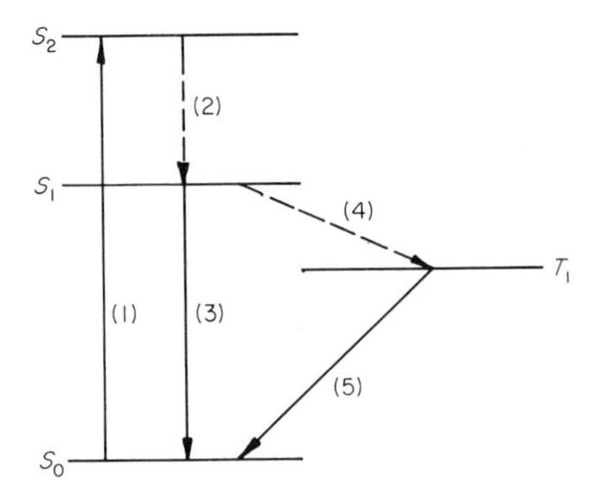

Fig. 4. Term scheme.

place by irradiation in the second absorption band. Since the radiationless transition to the ground state is, as we have seen, a rather rare event, and since moreover internal conversion is more frequent by a factor of about 10^6 than intersystem crossing, the radiationless occupation of S_1 will follow as the favored process from S_2 (path 2). The same thing is true if excitation is to still higher singlet states, S_3, S_4, etc., which are not shown in Fig. 4. Starting from S_1 two processes are possible: radiant transition to the ground state (path 3) and intersystem crossing to the lowest triplet state T_1 (path 4). The first transition gives, in combination with nuclear vibrations of the ground state, the *fluorescence spectrum* of our compound. The rate k_f of the fluorescence transition is proportional to the integrated intensity of the $S_0 \rightarrow S_1$ absorption transition. Assuming that the $S_0 \rightarrow S_1$ band with which we are concerned is one of medium intensity, such as the α band of an aromatic hydrocarbon, we can take k_f to be 10^8 sec^{-1}. With $k_{is} = 10^7$ sec^{-1} for the intersystem crossing it appears, then, that some 10% of the molecules present in the state S_1 reach the lowest triplet state T_1 by intersystem crossing, the greater part of the excited molecules going by fluorescence directly to the ground state. The reason for this, that the triplet state is relatively highly populated, is to be found in the fact that generally radiant transitions take place much more slowly than radiationless transitions; hence fluorescence transitions permitted by multiplicity and intersystem crossing become competitive. It may thus be shown that the occupation of the lowest triplet state by intersystem crossing is ca. 10^5 times as efficacious as that resulting from direct absorption of radiation.

The transfer of radiation from the lowest triplet state T_1 to the ground state (path 5 in Fig. 4) is forbidden within the limits of the previous discussion. It therefore takes place with a relatively long decay period and is observed as *phosphorescence*. This decay period of the radiation, its *lifetime*, depends very much on the type of compound, its environment, and especially the extent to which internal or external influences induce spin–orbit coupling. The lifetime can therefore vary over a wide range from ca. 10^{-4} second to several seconds. In liquid solutions and at temperatures around 20°C there occur bi- and monomolecular quenching processes that lead to radiationless thermal deactivation of the triplet state at quite high speed. It is therefore to be expected that $T_1 \rightarrow S_0$ phosphorescence will only be observable in liquid solutions of compounds whose triplet states have relatively short lifetimes.

In such cases there is observed, beside the extremely rapidly decaying fluorescence, a rather more slowly decaying luminescence of longer wavelength. Lewis and Kasha[5] were the first to demonstrate the possibility of observing rapidly decaying phosphorescence in liquid solutions. A recent example comes from the work of Parker and Hatchard.[13] These authors observed in liquid solutions of eosin in glycerol or ethanol at room temperature as well as the fluorescence a long-wavelength phosphorescence with a lifetime of the order of several milliseconds (see also Section 1.6). It is not surprising that occasionally compounds with very short-lived triplet states also display phosphorescence in the gas phase. An example that has been studied very frequently is biacetyl.[5, 14]

[13] C. A. Parker and C. G. Hatchard, *Trans. Faraday Soc.* **57**, 1894 (1961).

[14] J. W. Sidman and D. S. McClure, *J. Am. Chem. Soc.* **77**, 6461 and 6471 (1955); J. Heicklen, *ibid.* **81**, 3863 (1959).

If one employs conditions such that quenching processes of the kind mentioned are effectively excluded, then phosphorescence ought to be observable from compounds with quite long-lived triplet states. The most important method consists of "freezing" the bi- or monomolecular processes. One works, therefore, with rigid solutions and at low temperatures. Usually the substance being investigated is dissolved in an organic solvent or a mixture of solvents that has the property of setting to a glass at low temperature. The medium most frequently used is a mixture of ethanol, ether, and isopentane (EPA[15]; see Section 3.2). As cooling agents, liquified gases such as nitrogen, air, and helium are generally employed. With these "rigid solvents" it has been possible to observe the phosphorescence of numerous substances, aromatic hydrocarbons, heterocyclics, and their substitution products preponderating. The numerous investigations that have been carried out in this field have shown that phosphorescence is a quite general property of unsaturated organic compounds. Along with the phosphorescence, fluorescence occurs in the majority of cases. As has been shown, the two kinds of luminescence differ considerably in their lifetimes and so they can be distinguished in a simple way by the use of a phosphoroscope.

For the development of phosphorescence spectroscopy the introduction of the rigid solvent by Schmidt in 1896[16] was of the greatest significance. It is therefore all the more surprising that the part the rigid solvent plays in the occurrence of phosphorescence is not yet fully understood. Thus Porter and Wright[17] found that the radiationless deactivation of the triplet state in liquid solutions of aromatic hydrocarbons, e.g., naphthalene, occurs only in part by a second-order process and preponderantly by a first-order reaction. The velocity constants of the two processes are highly sensitive to the viscosity of the solvent, and fall as this increases. Hence the interesting question arises: what is the nature of a radiationless deactivation process that on the one hand follows first-order kinetics and on the other is dependent on viscosity? Porter and Wright suppose that when a molecule is in the triplet state its nuclear configuration is quite different from that in the ground state; in rigid solvents the triplet configuration is "frozen" so that a radiationless transition to the ground state is impossible. With falling viscosity

[15] G. N. Lewis and M. Kasha, *J. Am. Chem. Soc.* **66**, 2100 (1944).
[16] G. C. Schmidt, *Ann. Physik* [3] **58**, 103 (1896).
[17] G. Porter and M. R. Wright, *Discussions Faraday Soc.* **27**, 18 (1959).

the constraint on the nuclear configuration is relieved and the frequency of the radiationless deactivation of the triplet state increases proportionately.

The phenomenon of the phosphorescence of organic compounds has been known for quite a long time. The first observation would seem to have been made by Wiedemann,[18] who in 1888 demonstrated long-lived photoluminescence by exposing solid solutions of a number of organic dyes. Later Jablonski, Kautsky, Kowalski, Tiede, Schmidt, and others made important contributions to the subject.[19] The triplet theory of the phosphorescence of organic compounds given above was established by Lewis and Kasha in 1944[15] following the preparatory work of Lewis *et al.*[20] and of Terenin.[21] The foundations of the theory can now be regarded as firmly established. In particular the triplet nature of the "phosphorescent state" of organic compounds has been definitely verified in a series of experimental and theoretical investigations.

Conclusive experimental proof has been given that phosphorescent solutions show the paramagnetism to be expected from the participation of a triplet state. Thus Lewis and co-workers[22] found that solid solutions of fluorescein in boric acid had, under conditions suitable for phosphorescence, a paramagnetic susceptibility approximately equal to that expected for a triplet state. Evans[23] was able to demonstrate even more convincingly the relationship between phosphorescence and the triplet state that its paramagnetism indicated. He proved that in solid solutions of triphenylene in boric acid at room temperature the phosphorescence of triphenylene and its paramagnetism both fade away according to the same rate law. Both Lewis *et al.* and Evans employed the method of the "magnetic balance" for their measurements. Evidence for the paramagnetism of phosphorescent solutions has only recently become available from electron spin resonance (ESR) measurements because of

[18] E. Wiedemann, *Ann. Physik* [3] **34**, 446 (1888).

[19] See Th. Förster, "Fluoreszenz organischer Verbindungen," p. 261ff. Vandenhoeck & Ruprecht, Göttingen, 1951; P. Pringsheim, "Fluorescence and Phosphorescence," p. 285ff. Wiley (Interscience), New York, 1963.

[20] G. N. Lewis, D. Lipkin, and T. T. Magel, *J. Am. Chem. Soc.* **63**, 3005 (1941).

[21] A. Terenin, *Acta Physicochim. URSS* **18**, 210 (1943).

[22] G. N. Lewis and M. Calvin, *J. Am. Chem. Soc.* **67**, 1232 (1945); G. N. Lewis, M. Calvin, and M. Kasha, *J. Chem. Phys.* **17**, 804 (1949).

[23] D. F. Evans, *Nature* **176**, 777 (1955).

a series of experimental difficulties. Hutchison and Mangum[24] were the first who obtained an unequivocally interpretable ESR spectrum by exposing to ultraviolet (UV) light single crystals of durol containing naphthalene. Subsequently a whole series of similar studies has appeared, including aromatic hydrocarbons such as phenanthrene,[25] triphenylene,[26] coronene,[26,27] chrysene,[27,28] fluoranthene,[27] pyrene,[28] pyrene-d_{10},[29] 1,2-benzanthracene,[28] etc., heterocyclics such as quinoxaline,[30] quinoline,[31] isoquinoline,[31] etc., and derivatives of acetophenone.[32]

An indirect proof of the validity of the triplet theory of phosphorescence has been furnished by McClure,[3] who demonstrated that the phosphorescence shows the behavior to be anticipated for an intercombination transition. Mention has already been made of the fact, known from atomic spectroscopy, that in progressing from light to heavy elements the probability of intercombinations increases greatly because of increasing spin–orbit coupling. In molecules the extent of the spin–orbit coupling is determined by the heaviest atom that is present. If the phosphorescence of organic molecules corresponds to an intercombination transition, then the introduction of heavier substituents such as bromine or iodine into, say, aromatic hydrocarbons ought to increase markedly the probability of the phosphorescence transition, and this must be displayed as a reduction in the phosphorescent lifetime. This is exactly what McClure[3] observed with naphthalene and its halogen substitution products. The lifetime of the phosphorescence fell along the series naphthalene, fluoro-, chloro-, bromo-, iodonaphthalene.

[24] C. A. Hutchison and B. W. Mangum, *J. Chem. Phys.* **29**, 952 (1958); *ibid.* **34**, 908 (1961); see also A. Schmillen and G. von Foerster, *Z. Naturforsch.* **16a**, 320 (1961); A. W. Hornig and J. S. Hyde, *Mol. Phys.* **6**, 33 (1963).
[25] R. W. Brandon, R. E. Gerkin, and C. A. Hutchison, *J. Chem. Phys.* **41**, 3717 (1964).
[26] J. H. van der Waals and M. S. de Groot, *Mol. Phys.* **2**, 333 (1959); M. S. de Groot and J. H. van der Waals, *ibid.* **3**, 190 (1960).
[27] G. von Foerster, *Z. Naturforsch.* **18a**, 620 (1963).
[28] A. K. Piskunov, R. N. Nurmukhametov, D. N. Shigorin, V. J. Muromtsev, and G. A. Ozerova, *Izv. Akad. Nauk SSSR, Ser. Fiz.* **27**, 634 (1963); *Chem. Abstr.* **59**, 7085 (1963).
[29] S. W. Charles, P. H. H. Fischer, and C. A. McDowell, *Mol. Phys.* **9**, 517 (1965).
[30] J. S. Vincent and A. H. Maki, *J. Chem. Phys.* **39**, 3088 (1963).
[31] J. S. Vincent and A. H. Maki, *J. Chem. Phys.* **42**, 865 (1965).
[32] L. H. Piette, J. H. Sharp, T. Kuwana, and J. N. Pitts, *J. Chem. Phys.* **36**, 3094 (1962).

The corresponding results for the series of 1-halogenonaphthalenes are assembled in Table 1. McClure was able to show that the relationship observed between duration of phosphorescence and the atomic numbers of the substituents can be understood quantitatively on the basis of the triplet theory of phosphorescence. More will be said later in various

TABLE 1 Lifetimes of Phosphorescence in 1-Halogenonaphthalenes[a]

Compound	Phosphorescent lifetime (sec.)
Naphthalene	2.6
1-Fluoronaphthalene	1.5
1-Chloronaphthalene	0.3
1-Bromonaphthalene	0.018
1-Iodonaphthalene	0.0025

[a] D. S. McClure, *J. Chem. Phys.* **17**, 905 (1949).

places about the effects of spin–orbit coupling on triplet–singlet transitions (see Section 1.2).

Quantum mechanical calculations supplied a further proof of the triplet theory of phosphorescence. Using various approximate methods, the energy of the lowest triplet state (compared with the ground state as zero) has been calculated for numerous organic compounds. These values agree (within limits appropriate to the approximation) with the measured positions of the phosphorescence transitions. In Table 2 the calculated and observed values are collected for a series of polycyclic aromatic hydrocarbons. These show in an entirely independent way that the phosphorescence must be concerned with the transition from the lowest triplet state of the molecule to the ground state.

That the interpretations of phosphorescence and of the weak long-wave absorption bands discussed above are consistent follows from a series of arguments. The 0,0 bands of the *T–S* absorption spectra are related as object and mirror image to the 0,0 bands of the phosphorescence spectra of the compounds. It is therefore obvious that the same excited state participates in both spectra. This also follows from the fact that the lifetime of the triplet state evaluated from the integrated intensities of the T–S absorption spectrum agrees in order of magnitude with the decay period of the phosphorescence,[3,7] though perfect agreement is not to be expected for various reasons (see Section 1.4). It can also be demonstrated conclusively that the strengths of T–S absorption

TABLE 2 Comparison of Measured and Quantum Mechanically Calculated Energies (in Wave Numbers) of the Lowest Triplet States of Some Aromatic Hydrocarbons

Hydrocarbon	E_T (meas.)	E_T (calc.)	Method of calculation
Naphthalene	21,246	22,400	FE[a]
		22,700	LCAO empirical[b]
Anthracene	14,927	14,700	FE[a]
		15,200	LCAO empirical[b]
		14,000	ASMO-LCAO[c]
Tetracene	10,250	10,700	FE[a]
		10,800	LCAO empirical[b]
		10,000	Moffitt's method[d]
Phenanthrene	21,600	22,900	FE[a]
		22,200	LCAO empirical[b]
Coronene	19,410	22,000	FE[a]

[a] FE = free electron approximation. N. S. Ham and K. Ruedenberg, *J. Chem. Phys*. **25**, 13 (1956).
[b] G. G. Hall, *Proc. Roy. Soc.* **A213**, 113 (1952).
[c] R. Pariser, *Symp. Mol. Struct. Spectra, Ohio State Univ.* 1954.
[d] L. Goodman using the method of W. Moffitt, *J. Chem. Phys.* **22**, 320 (1954).

spectra are very sensitive to influences that affect spin–orbit coupling. The transition must, consequently, be recognized as an intercombination transition.

Every π-electron system exhibits, beside the lowest singlet and triplet states, a series of higher excited states of both multiplicities. It might be imagined that the occupation of a *higher* singlet excited state (by irradiation in the appropriate absorption band) would be followed by intersystem crossing to a higher triplet state and that from this, by a radiant transition, the molecule would descend to the ground state. The result would be that a substance would have several differently located phosphorescence spectra—and in exactly the same way several fluorescence spectra also. Experience shows, however, that this is not so, as can be easily understood since it has already been indicated that the process of internal conversion is quicker than other conceivable radiating or nonradiating transitions. The experimental fact that every substance displays only one fluorescence spectrum and only one phosphorescence spectrum has been summarized by Kasha[12] in the rule: "Only the lowest excited state of a given multiplicity is capable of emission," though he implies that the rule still requires considerable careful experimental and theoretical study "instead of being taken for granted as is commonly done."[33] All the same the majority of the exceptions to Kasha's rule so far discovered may well be simply the effects of impurities. An example may be discussed more thoroughly. Khaluporskii[34] reported that phenanthrene in ether at −180°C shows two phosphorescences that differ in their spectral positions and lifetimes. The longer-wave phosphorescence of the hydrocarbon this author observed agrees both in spectrum and in lifetime with the phosphorescence of phenanthrene as repeatedly described in the literature. The shorter-wave phosphorescence, which should also arise from phenanthrene, exhibits bands at about 411, 422, and 439 mμ and a lifetime of 1.6 second. These figures agree excellently with those of diphenylene sulfide (412, 425, and 440 mμ; lifetime 1.4 second). Diphenylene sulfide is well known as an impurity in phenanthrene isolated from bituminous coal tar. Thus what is called the "short-wave" phosphorescence of phenanthrene receives a simple explanation.[35] (Several phosphorescence spectra have also been described for carbazole and esculin,[36] and it cannot be seriously doubted that here too an impurity is responsible.) Ferguson and Tinson[37] noticed two phosphorescences in solid solutions of benzophenone in light petroleum at 77°K. The blue phosphorescence of dilute solutions

[33] M. Kasha, *Radiation Res.* Suppl. 2, 243 (1960).

[34] M. D. Khaluporskii, *Opt. i Spektroskopiya* **11**, 617 (1961); *Chem. Abstr.* **56**, 9589b (1962).

[35] M. Zander, unpublished data (1966).

[36] W. A. Pilipowitsch and B. Ja. Sweshnikow, *Ber. Akad. Wiss. Ud. SSR* [N.S.] **119**, 59 (1958).

[37] J. Ferguson and H. J. Tinson, *J. Chem. Soc.* p. 3083 (1952).

coincides with that recorded by other authors[15] for benzophenone. The green phosphorescence found in concentrated solutions (ca. 10^{-2} M) must, as Terenin and Ermolaev[38] have presumed, be attributed to the separation of crystalline benzophenone from the solution. Here too there is, apparently, no real exception to Kasha's rule. Special circumstances, however, seem to prevail at very low temperatures with crystals of aromatic hydrocarbons. Thus benzene crystals manifestly display two $T \rightarrow S$ luminescences[39] of which one could correspond to the transition from the lowest and the other to that from a higher triplet state to the ground state. On the other hand, this very same luminescence of crystals is extremely sensitive to impurities. With the quinones of the higher acenes[40] (tetracene, pentacene) two phosphorescences were definitely not observed, but it seems to be the case that the observed phosphorescence of acene quinones does not arise from the lowest triplet state but from a higher one and to that extent we seem here to have an exception to Kasha's rule (see Section 2.3). At any rate, for two compounds—azulene[41] and triphenylene[42]—there are hints of triplet–triplet fluorescence (i.e., of a radiating transition from an excited triplet to the lowest triplet state). Should these results be confirmed then we should have here a genuine exception to Kasha's rule; i.e., the luminescence does not have its origin in the lowest triplet state. There is an example of this type of emission experimentally confirmed; the singlet–singlet fluorescence of azulene[43] corresponds unequivocally to the transition from the *second* singlet excited state of the molecule to the ground state.

Although investigation in rigid solvents (including mixed crystals) is definitely the most generally useful method of studying phosphorescence, the phenomenon can also be observed occasionally in quite different conditions. Reference has already been made to phosphorescence in liquid solutions as well as in the gas phase. Publications concerning the phosphorescence of pure compounds in the crystalline state have also

[38] A. Terenin and V. Ermolaev, *Trans. Faraday Soc.* **52**, 1042 (1956).

[39] P. Pesteil, A. Zmerli, and L. Pesteil, *Compt. Rend.* **241**, 29 (1955).

[40] M. Zander, *Naturwissenschaften* **53**, 404 (1966).

[41] G. W. Robinson and R. P. Frosch, *J. Chem. Phys.* **38**, 1187 (1963); but see S. K. Lower and M. A. El-Sayed, *Chem. Rev.* **66**, 199 (1966).

[42] F. Dupuy, G. Nouchi, and A. Rousset, *Compt. Rend.* **256**, 2976 (1963).

[43] M. Beer and H. C. Longuet-Higgins, *J. Chem. Phys.* **23**, 1390 (1955); G. Viswanath and M. Kasha, *ibid.* **24**, 574 (1956).

appeared quite frequently. Robinson *et al.*[44] have tackled the problems of crystal phosphorescence with special thoroughness in the past few years. They have shown that crystal phosphorescence of absolutely pure compounds should be an extremely rare phenomenon, particularly for two reasons: Triplet excitation energies become delocalized extraordinarily quickly (more quickly than singlet excitation energies!); and they can be converted extremely effectively into singlet excitation energies by what is called triplet–triplet annihilation (see Section 1.6 for more about this). He and his school have estimated quantum yields of the order of magnitude of 10^{-9} for crystal phosphorescence on the basis of theoretical ideas they have developed; such weak luminescence cannot possibly be observed.

One suspects that much of the information contributed to the literature on crystal phosphorescence is to be attributed to chemical impurities, which need be present only in the smallest concentrations. In special cases, however, it ought to be possible to record "genuine" phosphorescence, that is, the $T \rightarrow S$ emission of the pure compound in the crystalline state. In particular this ought to be possible for substances whose phosphorescence is extraordinarily short lived, so that this transition can compete with the very rapid processes of triplet energy transfer and triplet–triplet annihilation. 2-Bromonaphthalene and 2-iodonaphthalene have extraordinarily short periods of phosphorescence. Sidman[45] has measured the crystal phosphorescence spectra of these compounds at 20°K and they clearly correspond to the T–S transitions of the pure compounds. They have been displaced toward lower wavelength by only about 500 cm^{-1} compared with the spectra in solution. The very large red shifts, which have been occasionally observed with other compounds when comparing results for solutions with those for crystals,[46] lead one to doubt the authenticity of the crystal phosphorescence spectra. Favorable conditions for the observation of "genuine" crystal phosphorescences, however, prevail with certain carbonyl compounds. These compounds generally have very short phosphorescent lives. In addition the phosphorescence is almost exclusively localized in the carbonyl group (see Section 2.2) and the carbonyl groups of neighboring molecules are relatively distant from each other in the crystal.

[44] H. Sternlicht, G. C. Nieman, and G. W. Robinson, *J. Chem. Phys.* **38**, 1326 (1963).
[45] J. W. Sidman, *J. Chem. Phys.* **25**, 229 (1956).
[46] See, for example, J. Czekalla and K. J. Mager, *Z. Elektrochem.* **66**, 65 (1962).

In this way the opportunities for rapid triplet energy transfer are limited. An example of this kind is benzophenone, which exhibits an easily observable *T–S* emission in the crystalline state.[47]

The observation of "genuine" crystal phosphorescences is undoubtedly favored by the use of very low temperatures (4°K). Measurements of this kind have been carried out by Pesteil *et al.*[48] in a series of investigations and by Kanda and Sponer.[49] That the phosphorescence of triphenylene crystals moves to longer wavelengths as the temperature rises (green phosphorescence at 77°K, yellow at 195°K, and red at 300°K) as observed by Zander,[50] is certainly the effect of impurity. This state of affairs ought to be explicable by the presence of different slight impurities, but about this nothing further will be said here. The phosphorescence of triphenylene in the crystalline state that Hochstrasser[51] has reported, and which only partially agrees with Zander's measurement at 77°K, results from impurities.[52] Finally it should be emphasized that much experimental and theoretical work will be necessary to bring nearer a clarification of the complicated problems of the phosphorescence of crystals.

1.2. Effects of Spin–Orbit Coupling

As we have already mentioned, spin–orbit coupling, i.e., the coupling of the orbital and the spin motions of individual electrons, causes perturbation of the electronic states by modification of their multiplicities. A triplet state thus acquires some singlet character as a result of "mixing" with a singlet state and, conversely, a singlet state acquires some triplet character. The probability of intercombination transitions is all the higher the greater this perturbation is, that is, the stronger the spin–orbit coupling in the atom or molecule. This applies to organic compounds

[47] D. S. McClure and P. L. Hanst, *J. Chem. Phys.* **23**, 1772 (1955).

[48] P. Pesteil, A. Zmerli, and L. Pesteil, *Compt. Rend.* **240**, 2217 (1955); P. Pesteil and M. Barbaron, *J. Phys. Radium* **15**, 92 (1954); P. Pesteil and A. Zmerli, *Ann. Phys. (Paris)* [12], 1079 (1955); *Cahiers Phys.* **55**, 71 (1956); *Compt. Rend.* **242**, 1876 (1956); L. Pesteil, P. Pesteil, and A. Zmerli, *ibid.* p. 2822; P. Pesteil and A. Zmerli, *ibid.* **243**, 1757 (1956); A. Zmerli, *J. Chim. Phys.* **56**, 405 (1959).

[49] Y. Kanda and H. Sponer, *J. Chem. Phys.* **28**, 798 (1958).

[50] M. Zander, *Naturwissenchaften* **49**, 7 (1962).

[51] R. M. Hochstrasser, *Rev. Mod. Phys.* **34**, 531 (1962).

[52] R. M. Hochstrasser, private communication (1963).

just as much for radiating transitions to the ground state (phosphorescence) as for radiationless transitions.

For light elements spin–orbit coupling is small. This is the case with organic compounds that are composed only of elements like carbon and hydrogen. On the other hand, it has long been known that intercombination transitions take place with considerable intensity in heavy atoms. The reason is that the spin–orbit coupling increases greatly under the influence of inhomogeneous electric fields such as are present in heavy atoms, i.e., in atoms of high atomic number. An entirely analogous effect is observed with organic compounds and is usually described concisely as the "heavy atom effect." Inhomogeneous magnetic fields, such as we have in paramagnetic elements, also increase the spin–orbit coupling and this magnetic effect, too, has been confirmed in organic compounds. Finally, spin–orbit coupling in organic molecules is enhanced through the formation of complexes by charge transfer. All three effects may be superposed in any particular case, though usually one will be dominant.

It is desirable that we distinguish between "internal" and "external" coupling effects. The term "internal" is used if the perturbing atom is a constituent element of the molecule whose T–S transitions are being investigated; in external effects it is a constituent either of the solvent or of some third component present in the solution.

Formation of complexes by charge transfer seems always to play some part in external spin–orbit coupling effects, and where the formation of such complexes is substantial it is the preponderant effect. Spin–orbit coupling effects are observable in rigid as well as in liquid solutions.

With increasing spin–orbit coupling in an organic molecule the probabilities of intersystem crossing ($S_1 \dashrightarrow T_1$), of phosphorescence transitions ($T_1 \rightarrow S_0$), and of radiationless transitions to the ground state ($T_1 \dashrightarrow S_0$) are all increased. Which process is influenced most must be determined in each particular case. Increased probability of the different processes is revealed in different ways:

(1) Enhancement of the $S_1 \dashrightarrow T_1$ transition causes increase of the quantum yield ϕ_p of the phosphorescence and, therefore, decrease of the quantum yield ϕ_f of the fluorescence and increase of the ratio ϕ_p/ϕ_f.

(2) Enhancement of the $T_1 \rightarrow S_0$ transition leads to decrease in the duration of phosphorescence and increase in the strength of the triplet–singlet absorption.

(3) Increase of the radiationless $T_1 \dashrightarrow S_0$ transition causes decrease of both the quantum yield and the duration of the phosphorescence. Numerous studies of these effects have appeared. A few may be discussed in detail.

The marked fall in the lifetime of the phosphorescence of naphthalene under the influence of heavy substituents such as bromine and iodine has already been mentioned (see Section 1.1).[1] It corresponds to an increase in the integrated intensity of the triplet–singlet absorption spectrum. McClure *et al.*[2] have shown that the observed changes can also be explained in an approximately quantitative manner on the basis of a spin–orbit coupling model. In this way it is found that the ratio of the integrated intensities of the *S–T* absorption spectra of 2-bromo- and 2-iodonaphthalene at about 1/5 agrees quite accurately with the ratio of the atomic spin–orbit coupling parameters (1/4.2) for bromine and iodine.

La Paglia[3] had investigated the phosphorescence and the *S–T* absorption of the tetraphenyl compounds of silicon, germanium, tin, and lead. As the atomic number, Z, of the metal atom increases the decay time of the phosphorescence falls and the intensity of the *S–T* absorption increases rapidly. Lead tetraphenyl ($Z_{Pb} = 82$; cf. $Z_I = 53$) has the most intense *S–T* absorption that has yet been observed in any π-electron system. It should be pointed out that atoms of high atomic number have only a very small influence on the *position* of the singlet–triplet transitions.

From observations that Kasha[4] and Ermolaev *et al.*[5] have reported, it follows that introduction of heavy atoms into the molecule of naphthalene raises not only the probability of the transition between the lowest triplet state and the ground state but also that of intersystem crossing. The ratio of the quantum yield in phosphorescence to that in fluorescence increases markedly in the series naphthalene, 2-chloro-, 2-bromo-, 2-iodonaphthalene, although the total quantum yield alters

[1] D. S. McClure, *J. Chem. Phys.* **17**, 905 (1949).
[2] D. S. McClure, N. W. Blake, and P. L. Hanst, *J. Chem. Phys.* **22**, 255 (1954).
[3] S. R. La Paglia, *J. Mol. Spectry.* **7**, 427 (1961); *Spectrochim. Acta* **18**, 1295 (1962).
[4] M. Kasha, *Radiation Res.* Suppl. 2, 243 (1960).
[5] V. L. Ermolaev and K. K. Svitashev, *Opt. i Spektroskopiya* **7**, 664 (1959); *Chem. Abstr.* **54,** 10,507f (1960); V. L. Ermolaev, J. P. Kotlyav, and K. K. Svitashev, *Izv. Akad. Nauk SSSR, Ser. Fiz.* **24**, 492 (1960); *Chem. Abstr.* **54,** 21,999c (1960).

only slightly. An entirely similar effect was found by Yuster and Weissman[6] for the chelates of dibenzoylmethane with the tervalent ions of certain metals (Al, Sc, Y, Lu, Gd, and La). Depending on the central metal ion the ratio ϕ_p/ϕ_f of the quantum yields of the phosphorescence and the fluorescence was found to change as well as the lifetime of the chelate phosphorescence. The atomic numbers of the metals, the lifetimes, and ϕ_p/ϕ_f values are collected in Table 3. The change in value as

TABLE 3 Effect of Atomic Number on Phosphorescent Lifetime[a] of Chelated Dibenzoylmethane[b]

Metal	Atomic number	Phosphorescent lifetime (sec.)	ϕ_p/ϕ_f
Al	13	0.50	—
Sc	21	0.30	0.15
Y	39	0.24	0.43
Lu(Cp)	71	0.12	1.16
La	57	0.09	2.32
Gd	64	0.002	No fluorescence observable

[a] All measurements at 77°K in EPA except Gd chelate, which was in alcohol.
[b] P. Yuster and S. J. Weissman, *J. Chem. Phys.* **17**, 1182 (1949).

one passes from Al to La, i.e., with rising atomic number, corresponds to the change already discussed for the halogenonaphthalenes and can be traced back to the simultaneous increase in the probabilities of intersystem crossing and of phosphorescence. Another possible explanation has been discussed by Reid.[7] The shortening of the lifetime of the triplet state with increasing atomic number that has been attributed to spin–orbit coupling makes triplet states increasingly less sensitive to quenching processes; hence, independently of intersystem crossing, an increase of ϕ_p/ϕ_f should take place.

Gadolinium is the only one of the metals investigated by Yuster and Weissman[6] that is paramagnetic. It shows a specially great reduction

[6] P. Yuster and S. J. Weissman, *J. Chem. Phys.* **17**, 1182 (1949).
[7] C. Reid, *Quart. Rev.* (*London*) **12**, 205 (1958).

in the lifetime of the chelate phosphorescence and, likewise, a great increase in ϕ_p/ϕ_f. It is obvious that here the electrical perturbation is superposed on a magnetic perturbation. Kasha and Becker[8] have described an interesting paramagnetic spin–orbit coupling effect of a similar kind. These authors found that the phthalocyanines of diamagnetic metal ions such as Mg(II) and Zn(II) show strong fluorescence while those of paramagnetic metal ions, like Ni(II), show phosphorescence but no fluorescence.

Many external spin–orbit coupling effects have been observed. One of these had long been known but was first correctly interpreted by Kasha,[9] viz., the quenching of the fluorescence of liquid solutions of aromatic hydrocarbons and other compounds by the presence of alkyl halides. This quenching of fluorescence increases with increasing atomic number of the halogen and is therefore to be recognized as typical of enhanced intersystem crossing arising from spin–orbit coupling. In rigid solutions at low temperature the suppression of fluorescence is linked with an increase of phosphorescence. McGlynn *et al.*[10] measured ϕ_p/ϕ_f values for rigid solutions of naphthalene in the presence of various alkyl halides. As the atomic number of the halogen gets larger an increase of ϕ_p/ϕ_f and a decrease of the duration of phosphorescence takes place. Which of the possible *S–T* intercombinations is the most sensitive to external spin–orbit coupling effects is not quite clear. McGlynn *et al.*[10] concluded, as a result of their measurements, that of the possible *S–T* intercombinations the transition probability of intersystem crossing is raised most. Siegel and Judeikis[11] concluded from measurements of phosphorescence and of ESR on systems quite similar to those investigated by McGlynn that the phosphorescence transition is more strongly influenced by external spin–orbit coupling than intersystem crossing. McGlynn *et al.*[12] have also investigated the halogenonaphthalenes as well as naphthalene. For the halogenonaphthalenes in the presence of alkyl halides, internal and external spin–orbit coupling effects are superposed. Various other derivatives of naphthalene such as the

[8] R. S. Becker and M. Kasha, *J. Am. Chem. Soc.* **77**, 3669 (1955).

[9] M. Kasha, *J. Chem. Phys.* **20**, 71 (1952); S. P. McGlynn, T. Azumi, and M. Kasha, *ibid.* **40**, 507 (1964).

[10] S. P. McGlynn, J. Daigre, and F. J. Smith, *J. Chem. Phys.* **39**, 675 (1963).

[11] S. Siegel and H. S. Judeikis, *J. Chem. Phys.* **42**, 3060 (1965).

[12] S. P. McGlynn, M. J. Reynolds, G. W. Daigre, and N. D. Christodouleas, *J. Phys. Chem.* **66**, 2499 (1962).

dinitronaphthalenes and compounds like coumarin and fluorescein also show more intense phosphorescence in the presence of ethyl iodide in rigid solvents, as was found by Graham-Bryce and Corkill.[13] Robinson *et al.*[14] have reported on a very interesting external spin–orbit coupling effect. They investigated the phosphorescence of benzene and deuterobenzene in rigid solutions of methane, argon, krypton, and xenon at 4.2°K. The pronounced fall in the lifetime of the phosphorescence of benzene or deuterobenzene between solution in argon (16 and 26 seconds, respectively) and solution in xenon (0.07 second for both compounds) points to a strongly Z-dependent spin–orbit coupling effect.

The strengths of singlet–triplet absorption spectra can also be altered by external spin–orbit coupling effects. Kasha[9] made the first observation of this kind. If the two colorless liquids 1-chloronaphthalene and ethyl iodide are brought together the mixture is yellow. Study of the visible spectrum reveals great similarity in position and vibrational structure to the commensurately strong S–T absorption spectrum of 2-iodonaphthalene. Obviously the weak S–T absorption of the 1-chloronaphthalene has been greatly enhanced under the influence of the ethyl iodide. McGlynn *et al.*[15] have established that we are definitely concerned here with an external spin–orbit coupling effect that increases the probability of the S–T transition. The effect is associated with the formation of a weak charge-transfer complex.

It has been observed by Evans[16–20] and other authors that S–T absorption spectra are greatly enhanced by the presence of paramagnetic substances such as oxygen, nitric oxide, or paramagnetic metal chelates[21] in the solvent. The effect had earlier been interpreted in rather obvious fashion as an overwhelming magnetic perturbation. Recent investigations[22] make it probable, however, that the increase in the intensity of

[13] J. J. Graham-Bryce and J. M. Corkill, *Nature* **186**, 965 (1960).

[14] M. R. Wright, R. P. Frosch, and G. W. Robinson, *J. Chem. Phys.* **33**, 934 (1960).

[15] S. P. McGlynn, R. Sunseri, and N. D. Christodouleas, *J. Chem. Phys.* **37**, 1818 (1962).

[16] D. F. Evans, *Nature* **178**, 534 (1956).

[17] D. F. Evans, *J. Chem. Soc.* p. 2753 (1959).

[18] D. F. Evans, *J. Chem. Soc.* p. 1351 (1957).

[19] D. F. Evans, *J. Chem. Soc.* p. 1753 (1960).

[20] D. F. Evans, *J. Chem. Soc.* p. 3885 (1957).

[21] J. N. Chaudhuri and S. Basu, *Trans. Faraday Soc.* **54**, 1605 (1958).

[22] H. Tsubomura and R. S. Mulliken, *J. Am. Chem. Soc.* **82**, 5966 (1960); J. N. Murrell, *Mol. Phys.* **3**, 319 (1960).

the *S–T* absorption spectra in these systems is connected chiefly with the formation of weak charge-transfer complexes. In these complexes the aromatic hydrocarbon or heterocyclic compound whose *S–T* absorption is being measured is behaving as an electron donor and the oxygen as an electron acceptor.

The method using oxygen is especially important and is carried out with liquid solutions (in special flasks) by keeping the oxygen in solution under high pressure. In this way *T–S* absorption spectra have been observed for numerous compounds, e.g., benzene,[16] benzene derivatives,[17] polycyclic aromatic hydrocarbons,[18] azahydrocarbons,[17] and also aliphatic compounds[19] such as ethylene, butadiene, and others. For easily volatilized substances, e.g., benzene, *T–S* absorption spectra have also been measured in the gaseous state by the oxygen method.[20]

Reference should also be made in this summary to the phosphorescent properties of *stable* charge-transfer complexes, e.g., those of polycyclic aromatic hydrocarbons (donors) with *sym*-trinitrobenzene or tetrachlorophthalic anhydride (acceptors). The phosphorescence of the donors is observed. The spectra are altered only slightly from those of the complex-free compounds and the duration of phosphorescence is abbreviated by the formation of the complexes. The work of Czekalla *et al.*[23] and of McGlynn *et al.*[24] has been specially important in elucidating for charge-transfer complexes the complicated problems that arise because the donor phosphorescence is superimposed on a charge-transfer fluorescence lying in the same spectral region.

By application of the external spin–orbit coupling effects just described, particularly Kasha's ethyl iodide method and Evans's oxygen method, *S–T* absorption spectra have now been made accessible for very many compounds. This is especially significant in cases in which the observation of the lowest triplet state by phosphorescence is impossible for experimental reasons such as very weak emission at very long wavelength. An example is tetracene, whose lowest triplet state has so far been

[23] J. Czekalla, *Naturwissenschaften* **43**, 467 (1956); J. Czekalla, G. Briegleb, W. Herre, and R. Glier, *Z. Elektrochem.* **61**, 537 (1957); J. Czekalla, A. Schmillen, and K. J. Mager, *ibid.* p. 1053; J. Czekalla, G. Briegleb, W. Herre, and H. J. Vahlsensiek, *ibid.* **63**, 715 (1959); J. Czekalla and K. J. Mager, *ibid.* **66**, 65 (1962).

[24] S. P. McGlynn and J. D. Bogus, *J. Am. Chem. Soc.* **80**, 5096 (1958); S. P. McGlynn, J. D. Bogus, and E. Elder, *J. Chem. Phys.* **32**, 357 (1960); for a comprehensive review, see S. P. McGlynn, *Chem. Rev.* **58**, 1113 (1958).

measurable only in absorption and not in emission.[25] Knowledge of the position of the lowest triplet state of this hydrocarbon has been of great significance for the theory (see Section 2.1).

External spin–orbit coupling effects can also find useful applications in phosphorimetry; details of these will be given later (see Section 3.2).

1.3. Positions and Vibrational Structures of Phosphorescence Spectra

If the simple Hückel molecular orbital approximation is used to derive the energies of singlet and triplet states having the same electronic configurations, i.e., having the same orbital distribution of electrons, it is found that they, the energies, are identical. The reason is that the coulombic repulsion of the electrons has not been allowed for. This repulsion will be all the greater and the energy of the excited state simultaneously all the higher, the smaller the distances between the various electrons. It can be shown that the probability of both electrons of a singlet state being found in the same place has a finite value, but that for the electrons of a triplet state is zero. Because the closer proximity of the electrons of the singlet state leads to greater coulombic repulsion, the singlet state always possesses higher energy than the triplet state with the same electronic configuration.[1] This is a special case of Hund's rule, applicable both to atoms and to molecules, that the state of highest multiplicity is always the most stable.

It follows from this that the lowest excited state of a molecule is always a triplet. Since the phosphorescence originates from this, it must necessarily lie at longer wavelength than the fluorescence that originates from the lowest singlet excited state.

The position of the phosphorescence transition depends strongly on the structure of the molecule, and the spectral region in which it may be observed stretches from the UV to the near infrared. Spectra lying in the visible region are obviously very easily accessible experimentally, whereas those at very long wavelength, which are usually also extremely weak, can only be measured with difficulty.

[25] S. P. McGlynn, M. R. Padhye, and M. Kasha, *J. Chem. Phys.* **23**, 593 (1955).

[1] A. Streitwieser, Jr., "Molecular Orbital Theory for Organic Chemists," Wiley, New York, 1961; C. Sandorfy, "Electronic Spectra and Quantum Chemistry." Prentice-Hall, Princeton, New Jersey, 1964.

The phosphorescence spectra of the aromatic hydrocarbons, of most heterocyclics, and of many derivatives of these classes of compounds correspond to π–π^* transitions, but although this type of excitation is by far the most frequent, there is a series of compounds that shows n–π^* phosphorescence.[2] To this belong very important families of compounds: the carbonyls (quinones, ketones, etc.), the nitro compounds, and a few heterocyclics such as pyrimidine and pyrazine.

A series of criteria[2] is available by which π–π^* and n–π^* phosphorescences can be distinguished. The latter are characterized usually by a very high ratio of ϕ_p/ϕ_f. Thus, for example, anthraquinone (see Section 2.3) shows no measurable fluorescence, but instead, an intense n–π^* phosphorescence; anthracene fluoresces more strongly than it phosphoresces.

In homologous chemical series, the energies and lifetimes of phosphorescence are strongly dependent on the size of the molecule if they result from π–π^* transitions, but are only slightly influenced by molecular size if they result from n–π^* transitions.

For azahydrocarbons, Goodman and Kasha[3] found a sharp criterion for distinguishing the two types of phosphorescence, viz., that π–π^* phosphorescence is displaced toward longer wavelengths by substitution, for example, by methyl groups; n–π^* phosphorescence, on the other hand, is displaced toward shorter wavelengths. Goodman and Shull[4] have also demonstrated a theoretical foundation for this criterion.

Finally it should be mentioned that in the vibrational structure of n–π^* phosphorescence spectra there are usually to be found frequencies characteristic of those groups present that have lone electron pairs, for example, the carbonyl frequency (ca. 1725 cm^{-1}) in the phosphorescence spectra of carbonyl compounds and the Raman frequency of the nitro group (ca. 1450 cm^{-1}) in the phosphorescence spectra of aromatic nitro compounds.[2]

The band of shortest wavelength (the 0,0 band) of a phosphorescence spectrum gives the energy (in cm^{-1}) of the lowest triplet state compared with the ground state as the zero of energy. In correlating the triplet state with the singlet excited states of the molecule, three questions are especially interesting:

[2] M. Kasha, *Radiation Res.* Suppl. 2, 243 (1960).
[3] S. L. Goodman and M. Kasha, *J. Mol. Spectry.* **2**, 58 (1958).
[4] S. L. Goodman and L. Shull, *J. Chem. Phys.* **27**, 1388 (1957).

1. How great is the energy difference between the lowest singlet and the lowest triplet state? This energy difference makes it possible to predict whether a thermal transition from the lowest triplet to the lowest singlet state is possible, and this is significant in connection with what is called "E-type delayed fluorescence" (see Section 1.6). The difference in energy is easily deduced from the phosphorescence and fluorescence spectra.

2. Which singlet excited state has the same electronic configuration as the lowest triplet state? This question concerns the classification[5] of the lowest triplet state in a compound and the consequent magnitude of the singlet–triplet splitting (the energy difference between the singlet and triplet states having the same electronic configurations). There are two important possibilities for the classification. First, in homologous chemical series (e.g., hydrocarbons of the acene series) passage from one member of the series to the next is accompanied by approximately uniform changes, both in direction and magnitude, in those spectra that arise from singlet and triplet states having the same electronic configuration. Second, classification frequently becomes feasible if comparison is made between the measured energy of the lowest triplet state of a molecule and the energies of successive triplet states as determined quantum mechanically.

The classification of the lowest triplet state is thoroughly discussed for polycyclic hydrocarbons in Section 2.1.

3. Which singlet state mixes with the lowest triplet state (see Section 1.4)? Since the polarization of the phosphorescence is determined by the symmetry of that singlet state that perturbs and mixes with the triplet state, conclusions can be reached from the phosphorescence polarization spectra concerning the nature of the perturbing singlet state. Such spectra have been measured for a large number of compounds and will be considered later (see Section 1.4).

In addition to the position of the 0,0 band of a phosphorescence spectrum, its vibrational structure is also of interest. Because of the low temperature at which phosphorescence measurements are usually carried out, the phosphorescence transition starts at the lowest vibrational level of the triplet state and leads into the various vibrational levels of the ground state. The distances between the phosphorescence 0,0 band and the higher bands are therefore given by the nuclear vibration

[5] D. R. Kearns, *J. Chem. Phys.* **36**, 1608 (1962).

frequencies of the ground state. The succession of the frequencies determined from the phosphorescence spectra agrees with data from infrared and Raman spectra.

For the detailed vibrational analysis, spectra that are rich in bands of small half-value width are desirable. In connection with the important fluorescence investigations of Shpol'skii,[6] Sponer, Kanda, and their collaborators[7] have shown that phosphorescence spectra that are especially rich in structural details are obtained by employing a crystalline hydrocarbon matrix. This has been demonstrated particularly well with cyclohexane.

Vibrational analyses of numerous phosphorescence spectra are to be found in the literature (see Section 2.1). For the overwhelming majority of compounds the vibrational structure and the intensity distribution of the spectra show that the phosphorescence transition is not forbidden by symmetry (concerning the selection rules in organic molecules, see Streitwieser[1]). The 0,0 bands of the spectra are intense and usually display totally symmetrical vibrations. Examples of typical phosphorescence spectra (phenanthrene and carbazole) are shown in Fig. 5.

An exception to the usual behavior is found in benzene. Working independently Shull[8] and Dikun and Sveshnikov[9] were able to show that the phosphorescence transition of benzene is symmetry forbidden. The triplet–singlet transition of this compound is therefore determined by *two* selection rules, viz. intercombination forbidden and symmetry forbidden. That the spectrum is symmetry forbidden is shown by the extremely low intensity of the 0,0 band (in EPA at 77°K); moreover it follows unequivocally from the vibrational analysis of the spectrum given by these authors.

Extremely weak 0,0 bands also distinguish the phosphorescence spectra of the hydrocarbons triphenylene and coronene which are also symmetrical; for these compounds too it seems probable that the phos-

[6] E. V. Shpol'skii, *Usp. Fiz. Nauk* **71**, 215 (1960); *ibid.* **80**, 255 (1963).

[7] H. Sponer, Y. Kanda, and L. A. Blackwell, *Spectrochim. Acta* **16**, 1135 (1960); Y. Kanda and R. Shimada, *ibid.* **17**, 279 (1961); Y. Kanda, R. Shimada, and Y. Sakai, *ibid.* **17**, 1 (1961); Y. Kanda, R. Shimada, K. Hanada, and S. Kajigaeshi, *ibid.* **17**, 1268 (1961); R. Shimada, *ibid.* **17**, 14 and 30 (1961).

[8] H. J. Shull, *J. Chem. Phys.* **17**, 295 (1949).

[9] P. P. Dikun and B. Y. Sveshnikov, *Zh. Eksperim. i. Teor. Fiz.* **19**, 1000 (1949); *Dokl. Akad. Nauk SSSR* **65**, 637 (1949).

phorescence transition is symmetry forbidden.[10]

Several observations suggest that the forbidding of phosphorescence transitions by symmetry may be slightly relaxed through external perturbations. Thus Kanda and Shimada[11] found that benzene in a crystalline matrix of carbon tetrachloride (in contrast to its properties in an EPA glass) displays a typical symmetry-permitted phosphorescence spectrum with an intense 0,0 band.

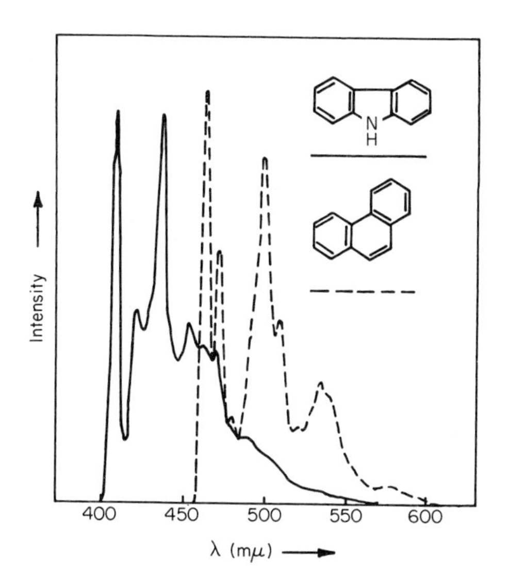

Fig. 5. Phosphorescence spectra of carbazole (—) and phenanthrene (--) in EPA at 77°K.

Note: In all figures showing spectra the intensities are in (uncorrected) scale divisions unless otherwise indicated.

Robinson *et al.*[12] have observed that in rigid solutions in krypton and xenon (solvents in which a considerable external spin–orbit coupling effect is revealed as a decrease in the duration of phosphorescence), benzene and deuterobenzene have intense phosphorescence 0,0 bands,

[10] F. Dörr and H. Gropper, *Z. Elektrochem.* **67**, 193 (1963); J. Czekalla and K. J. Mager, *ibid.* **66**, 65 (1962).

[11] Y. Kanda and R. Shimada, *Spectrochim. Acta* **17**, 7 (1961).

[12] M. R. Wright, R. P. Frosch, and G. W. Robinson, *J. Chem. Phys.* **33**, 934 (1960).

but in methane and argon (solvents in which no spin–orbit coupling effect occurs) the 0,0 bands are extremely weak.

Zander has made a very similar observation[13] on triphenylene and coronene. The phosphorescence 0,0 bands of these hydrocarbons are very weak in EPA as well as in crystalline matrices of *n*-heptane, benzene, toluene, or *p*-xylene (solvents without spin–orbit coupling effects),

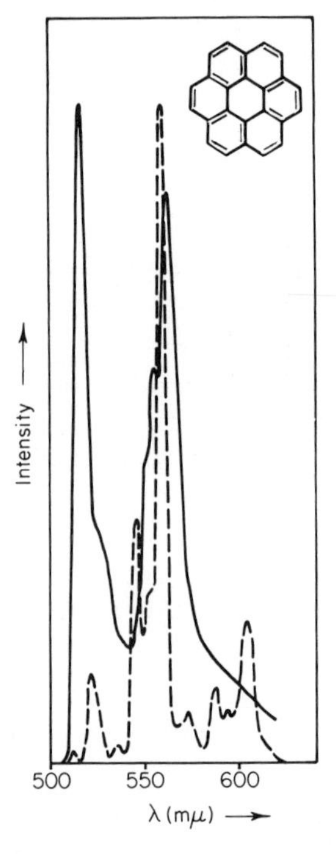

Fig. 6. Phosphorescence spectra of coronene in trichlorobenzene (—) and EPA (--) at 77°K. Intensity = relative energy (corrected).

though they are very intense in carbon tetrachloride, monochlorobenzene, *o*-dichlorobenzene, 1,2,4-trichlorobenzene, or monobromobenzene (solvents having spin–orbit coupling effects). The positions of the phosphorescence transitions were only slightly influenced by the rigid solvents. As an example, the phosphorescence spectra of coronene

[13] M. Zander, *Naturwissenschaften* **52**, 559 (1965).

in EPA and in 1,2,4-trichlorobenzene are repeated in Fig. 6. The following observation shows that this last effect is not attributable to the crystalline matrix: the 0,0 band of triphenylene or coronene, which is very weak in EPA, becomes very strong in a glass that consists of 9 parts of EPA and 1 part of methyl iodide (see also Section 3.2). Also the symmetry-forbidden *S–T* absorption spectrum of benzene becomes symmetry permitted[14] in solution and in the gas phase in the presence of oxygen (increase of spin–orbit coupling; see Section 1.2); this is demonstrated by the great increase in the intensity of the 0,0 band.

All these observations point to an incomplete understanding of the connection between the prohibition of phosphorescence by symmetry and that by intercombination; this is obvious from the symmetrical hydrocarbons discussed here.

Except in the cases just considered the positions and the vibrational structures of phosphorescence spectra alter only slightly with the solvent as has been shown especially by the extended researches of Nauman.[15]

Little is known about the dependence of these spectra on temperature. The spectrum of coronene in rigid solution in perhydrocoronene matches that in EPA at 77°K. At room temperature new bands appear at shorter wavelength and can perhaps be attributed to transitions from higher nuclear vibration terms of the triplet state.[16]

1.4. Lifetime, Quantum Yield, and Polarization of Phosphorescence

In rigid glasses and at low temperature, phosphorescence decays exponentially according to the first-order law. To determine the "mean phosphorescent lifetime" τ_0, the intensity of the phosphorescence is plotted logarithmically against the time after cutting off the exciting radiation. The gradient of the resulting straight line is $2.303\tau_0$.[1,2]

The quantity τ_0 is only identical with the "natural lifetime" τ_N of the triplet state if the sole transition arising from this state T_1 is the *radiation*

[14] D. F. Evans, *J. Chem. Soc.* p. 3885 (1957).

[15] R. V. Nauman, Dissertation, University of California, Berkeley, California (1947).

[16] M. Zander, *Naturwissenchaften* **47**, 443 (1960).

[1] See T. Förster, "Fluoreszenz organischer Verbindungen," pp. 156ff. and 266ff. Vandenhoeck & Ruprecht, Göttingen, 1951.

[2] D. S. McClure, *J. Chem. Phys.* **17**, 905 (1949).

transition to the ground state S_0. This is not often the case. Usually a radiationless transition to the ground state also occurs and has greater probability the smaller the energy difference between T_1 and S_0.[3] Because of this radiationless transition $T_1 \dashrightarrow S_0$, τ_0 is generally smaller than τ_N.

Assuming that the only radiationless transition taking place in a molecule occurs from T_1 to the ground state and not from the lowest excited singlet state or any higher excited state, it is possible to calculate τ_N from a knowledge of the absolute quantum yields of fluorescence and phosphorescence.[4] The relationship is

$$\tau_N = \tau_0(1 - \phi_f)/\phi_p \tag{2}$$

In cases in which other radiationless transitions to the ground state occur as well as $T_1 \dashrightarrow S_0$, the "true" τ_N is smaller than that given by Eq. (2). From this it follows that the true τ_N lies between τ_0 and that given by Eq. (2). The range of values obtained in this way is useful for many semiquantitative purposes.

For reasons that Robinson[5] (in particular) has discussed, the radiationless $T_1 \dashrightarrow S_0$ transition is almost completely suppressed in perdeuterated hydrocarbons (e.g., perdeuterobenzene). Hence the mean phosphorescent lifetime τ_0 of a perdeuterated compound is approximately the same as the natural lifetime τ_N of the triplet state.[5,6] The frequencies of other processes are not substantially altered for deuterated compared with nondeuterated compounds. The τ_0 values of the deuterated compounds are, therefore, simultaneously good approximations to the τ_N values of the nondeuterated compounds. In Table 4, τ_0 is listed for some deuterated and nondeuterated aromatic hydrocarbons.[7] The significantly smaller values of τ_0 for the hydrogenated compounds show that in the hydrocarbons the radiationless $T_1 \dashrightarrow S_0$ transition can have a really high probability.

In principal it should be possible to calculate the natural lifetime τ_N

[3] G. W. Robinson and R. P. Frosch, *J. Chem. Phys.* **38**, 1187 (1963).

[4] E. H. Gilmore, G. E. Gibson, and D. S. McClure, *J. Chem. Phys.* **20**, 829 (1952); for correction, see *ibid.* **23**, 399 (1955).

[5] G. W. Robinson, *J. Mol. Spectry.* **6**, 58 (1961).

[6] M. S. de Groot and J. H. van der Waals, *Mol. Phys.* **4**, 189 (1961); E. C. Lim and J. D. Laposa, *J. Chem. Phys.* **41**, 3257 (1964).

[7] R. E. Kellog and R. P. Schwenker, *J. Chem. Phys.* **41**, 2860 (1964).

of the triplet state from the integrated intensity of the *S–T* absorption spectrum. In practice there are many difficulties. Thus, it is very difficult (e.g., because of traces of oxygen) to observe the entirely unperturbed *S–T* spectrum. In any case entirely unperturbed *S–T* spectra are usually very weak and so in determining the area under the absorption curve through the spectral background originating from the longest-wavelength

TABLE 4 Phosphorescent Lifetimes of Perdeuterated Aromatic Hydrocarbons[a]

Compound	Phosphorescent lifetime (sec)[b]
Naphthalene-d_8	22.0
Naphthalene-h_8	2.4
Phenanthrene-d_{10}	16.4
Phenanthrene-h_{10}	3.8
Triphenylene-d_{12}	23.0
Triphenylene-h_{12}	16.0
Pyrene-d_{10}	3.2
Pyrene-h_{10}	0.5
Diphenyl-d_{10}	10.3
Diphenyl-h_{10}	4.2
p-Terphenyl-d_{14}	5.3
p-Terphenyl-h_{14}	2.6

[a] R. E. Kellog and R. P. Schwenker, *J. Chem. Phys.* **41**, 2860 (1964).
[b] All measurements in EPA at 77°K.

S–S band large errors may be introduced. Finally the relationship between the lifetime of the excited state and the absorption strength of the corresponding transition is only approximately valid for molecules.[4,8] All the same, in cases in which the triplet lifetime measured from phosphorescence is appreciably shorter than that determined from the *T–S* absorption, one can conclude that there is a high frequency of radiationless $T_1 \dashrightarrow S_0$ transitions.

[8] G. N. Lewis and M. Kasha, *J. Am. Chem. Soc.* **67**, 994 (1945); M. R. Padhye, S. P. McGlynn, and M. Kasha, *J. Chem. Phys.* **24**, 588 (1956).

TABLE 5 Phosphorescent Lifetimes of Various Compounds

Compound	Phosphorescent lifetime (sec.)	Reference[a]
	Aromatic Hydrocarbons	
Benzene	7.0	1
Naphthalene	2.6	1
Anthracene	0.09	2
Phenanthrene	3.3	1
1,2-Benzanthracene	0.3	1
Chrysene	2.5	1
3,4-Benzphenanthrene	3.5	3
Pyrene	0.2	1
1,2-Benzpyrene	2.0	3
1,2:6,7-Dibenzpyrene	7.5	3
Coronene	9.4	1
	Substituted Aromatics	
Chlorobenzene	0.004	1
Phenol	2.9	1
Aniline	4.7	1
Benzoic acid	2.5	1
1-Nitronaphthalene	0.05	1
1,5-Dinitronaphthalene	0.11	1
	Carbonyl Compounds	
Acetone	0.0006	1
Acetophenone	0.008	1
Anthraquinone	0.2	3
Methyl-2-naphthyl ketone	0.95	1
	Heterocyclics	
Indole	5.0	4
Quinoline	0.5	5
Isoquinoline	1.3	5
7,8-Benzoquinoline	1.4	5

[a] 1. D. S. McClure, *J. Chem. Phys.* **17**, 905 (1949).
2. S. P. McGlynn, M. R. Padhye, and M. Kasha, *J. Chem. Phys.* **23**, 593 (1955).
3. E. Clar and M. Zander, *Chem. Ber.* **89**, 749 (1956); M. Zander, unpublished data (1965).
4. R. C. Heckman, *J. Mol. Spectry.* **2**, 27 (1958).
5. D. P. Craig and J. G. Ross, *J. Chem. Soc.* p. 1589 (1954).

In Table 5 the mean phosphorescent lifetimes τ_0 of various compounds have been assembled. The measurements are for glassy rigid solvents at low temperatures. It will be seen that they are strongly dependent on the structures of the molecules and are quite short for carbonyl compounds. The influence of heavy substituents on the duration of phosphorescence has already been demonstrated in detail (see Section 1.1). For aromatic hydrocarbons a reduction of lifetime is observed with increasing displacement of the phosphorescence transition toward longer wavelengths. This is attributable to the rapid increase in the rate of the radiationless $T_1 \dashrightarrow S_0$ transition with decreasing difference between the T_1 and S_0 energies.

The mean phosphorescent lifetime τ_0 is effectively independent of concentration. Hence for a given temperature and solvent it is a characteristic constant of the substance and so can be used for its identification.[9] This has still to be discussed in detail (see Section 3.3.3).

The dependence of τ_0 on temperature is quite complicated and turns on the nature of the rigid solvent. For example, with aromatic hydrocarbons such as coronene, phenanthrene, fluoranthene, etc. in heptane it is only slightly temperature dependent. At about 170°K, i.e., at about 10°C below the melting point of the solvent, it has still about half the value measured at 77°K.[10] In artificial glasses such as polyacrylonitrile (Orlon A), cellulose acetate, and others the magnitudes of τ_0 for aromatic hydrocarbons again depend only slightly on temperature. Whatever the compound, τ_0 is reduced by a factor of about 2 between 77°K and room temperature.[7] For rigid solutions of naphthalene and of naphthalene-d_8 in durol[11] as well as for triphenylene and some other compounds in glycerol,[12] an identical temperature variation was observed, viz., slight dependence on temperature at low temperatures, strong dependence above 200°K. What kind of process comes into operation at the higher temperatures is not really understood; it leads to a very marked decrease of τ_0. In contrast, when these hydrocarbons are dissolved in isopropyl alcohol (again at very low temperatures), their τ_0 values display an extremely strong variation with temperature.[10, 13]

[9] M. Zander, *Angew. Chem. Intern. Ed. Engl.* **4**, 930 (1965).
[10] G. von Foerster, *Z. Naturforsch.* **18a**, 620 (1963).
[11] S. G. Hadley, H. E. Rast, and R. A. Keller, *J. Chem. Phys.* **39**, 705 (1963).
[12] T. H. Jones and R. Livingston, *Trans. Faraday Soc.* **60**, 2168 (1964).
[13] A. Schmillen and A. Tschampa, *Z. Naturforsch.* **19a**, 190 (1964).

The dependence of τ_0 on the solvent that is occasionally observed may be closely connected with its dependence on temperature.[10]

Absolute determinations of the quantum yields in phosphorescence are experimentally rather difficult to carry out and so have still not been made for very many compounds.[4, 14] A collection of values is shown in Table 6. As with the lifetime of phosphorescence so too its quantum

TABLE 6 Quantum Yields in the Phosphorescence of Various Compounds[a]

Compound	Quantum yield
Benzene	0.17
Naphthalene	0.06
Phenanthrene	0.31
Triphenylene	0.37, 0.46
Chrysene	0.13
Coronene	0.30
Chlorobenzene	0.04
Acetophenone	0.43
Benzophenone	0.52

[a] E. H. Gilmore, G. E. Gibson, and D. S. McClure, *J. Chem. Phys.* **20**, 829 (1952); W. H. Melhuish, *J. Opt. Soc. Am.* **54**, 183 (1964).

yield is very sensitive to molecular structure. It is found that the yield decreases with increasing shift of the transition toward longer wavelengths and this, like the parallel change in the lifetime, suggests a great increase of the radiationless $T_1 \dashrightarrow S_0$ transition. Since this process obviously cannot be neglected, particularly when the difference between S_0 and T_1 is small, it is only very approximately correct to equate the ratio of the quantum yields in fluorescence and phosphorescence with the ratio of the probabilities of fluorescence and intersystem crossing, as has occasionally been done.

The phosphorescence quantum yield is temperature dependent and, at least for aromatic hydrocarbons, more strongly so than the phos-

[14] W. H. Melhuish, *J. Opt. Soc. Am.* **54**, 183 (1964).

phorescent lifetime. It decreases rapidly with rising temperature, again over a range throughout which the medium is quite rigid.[10]

A matter of great interest is the question of whether the phosphorescence quantum yield and the ratio of the quantum yields in phosphorescence and fluorescence ϕ_p/ϕ_f depend on the exciting wavelength. For the majority of the compounds investigated it has been established that this is not so.[4,15] However, for some, e.g., fluorescein,[16] chrysene,[17] and the interesting substance hexahelicene[17] (I), ϕ_p/ϕ_f increases with

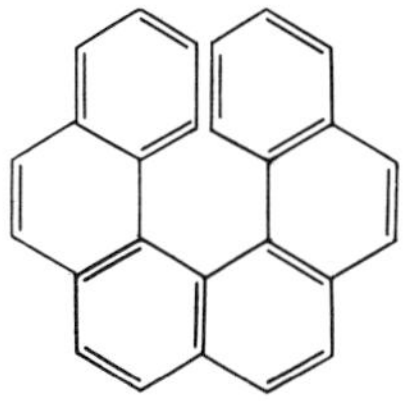

I

falling excitation wavelength. A possible explanation is that in these cases intersystem crossing from the higher singlet excited states occurs with greater frequency than from the lowest singlet excited state. It can be established theoretically that the opposite case cannot occur,[3] and it has, in fact, never been observed.

Ferrocene behaves in an interesting way. As Scott and Becker[18] found, no phosphorescence takes place following irradiation in the absorption band of longest wavelength (4300 Å); it is observed at 5280 Å, with $\tau_0 = 2$ seconds, however, in EPA at 77°K if irradiation is in the second absorption band (3240 Å), and it arises from the lowest triplet level of ferrocene. Several possible explanations of this unusual phosphorescence excitation phenomenon have been discussed.

According to theory the rule forbidding intercombination relaxes

[15] J. Ferguson, *J. Mol. Spectry.* **3**, 177 (1959); R. M. Hochstrasser, *Can. J. Chem.* **38**, 233 (1960).

[16] See P. Pringsheim, "Fluorescence and Phosphorescence," p. 437. Wiley (Interscience), New York, 1963; R. Bauer and A. Baczynzki, *Bull. Acad. Polon. Sci., Ser. Sci., Math., Astron. Phys.* **6**, 113 (1958); *Chem. Abstr.* **52**, 13,438 (1958).

[17] M. F. O'Dwyer, M. Ashraf El-Bayoumi, and S. J. Strickler, *J. Chem. Phys.* **36**, 1395 (1962); but see S. K. Lower and M. A. El-Sayed, *Chem. Rev.* **66**, 199 (1966).

[18] D. R. Scott and R. S. Becker, *J. Chem. Phys.* **35**, 516 (1961).

because spin–orbit coupling mixes some singlet character into the triplet state.* How, in any particular case, can a decision be reached concerning the nature of the perturbing singlet state? It can be shown that the *polarization*† of the phosphorescence transition is identical with that of an absorption transition that leads from the ground state of the molecule to the perturbing singlet excited state. In this way measurements of the polarization of the phosphorescence permit decisions concerning the nature of the perturbing singlet state and in this lies their great significance for the theory of phosphorescence.[19]

It is, however, conceivable *a priori* that not just one but several different singlet excited states of comparable importance could get mixed into the triplet state. These different perturbations would then bring about in a compound several triplet–singlet radiating transitions of the same *energy* but of *different* polarizations. In such a case only a slight degree of polarization (zero, weak positive, or weak negative) would be expected as the mean of the different polarizations of the individual T–S transitions.

Relationships of this kind were found by Dörr *et al.* as well as by El-Sayed and Pavlopoulos[20] for the halogenonaphthalenes (1- and 2-fluoro-, chloro-, bromo-, and iodonaphthalene). Naphthalene itself shows negative polarization of the phosphorescence and the observed degree of polarization reaches approximately the theoretical limiting value of $-1/3$. Hence it is concluded that only a single perturbing singlet state is involved in mixing with the triplet state. Halogen substitution makes the degree of polarization more positive and in fact all the more so the greater the atomic number Z of the halogen. In no case does it approach the limiting value, positive or negative, for polarization

* It is assumed here that the perturbation of the singlet ground state, through which this acquires some triplet character, is negligibly small. In many cases, however, this is only approximately true. In this connection, see L. Goodman and V. G. Krishna, *J. Chem. Phys.* **37**, 2721 (1962).

† For the theoretical foundations of the polarization of phosphorescence and fluorescence, see Th. Förster, "Fluoreszenz organischer Verbindungen," p. 160. Vandenhoek & Ruprecht, Göttingen, 1951; S. K. Lower and M. A. El-Sayed, *Chem. Rev.* **66**, 199 (1966); F. Dörr, *Angew. Chem.* **78**, 457 (1966).

[19] H. Gropper and F. Dörr, *Z. Elektrochem.* **67**, 46 (1963).

[20] F. Dörr, H. Gropper, and N. Mika, *Z. Naturforsch.* **18a**, 1025 (1963); M. A. El-Sayed and T. Pavlopoulos, *J. Chem. Phys.* **39**, 1899 (1963); *ibid.* **41**, 1082 (1964).

exclusively in one direction. It is, therefore, obviously the case here that not a single singlet excited state but several of different symmetry are mixed with the triplet state. It is, of course, plausible that the perturbation introduced into the molecule by substituents of high Z, e.g., iodine, is different from that in the hydrocarbon.

The properties observed for the halogenonaphthalenes are, however, relatively rare. In the majority of cases the same results are found as for naphthalene, namely, almost exclusive polarization of phosphorescence in one direction. In other words, only one perturbing singlet excited state participates.

It is very interesting to note that all the compounds with π–π^* phosphorescence that have been investigated in this connection show negative polarization whatever absorption band is selected for the excitation.[21–29] According to theory this means that the moment of the phosphorescence transition is perpendicular to the moments of all the absorption transitions appearing in the molecule. Since whenever these absorption transitions involve π–π^* transitions they are polarized *in* the molecular plane, the moment of the phosphorescence transition must be *perpendicular* to it.[21–29]

This result was obtained from a large number of compounds and although negative polarization of phosphorescence was first observed by Williams[21] for phenanthrene, McGlynn and Azumi,[22] among others, have since confirmed the discovery using more refined methods of measurement.

Krishna and Goodman[23] carried out measurements of the polarization of phosphorescence not only on phenanthrene but also on naphthalene, chrysene, and picene; in every case they found that the transition moment of the phosphorescence is approximately perpendicular to the plane of the molecule. Dörr and Gropper[24] came to the same conclusion

[21] R. Williams, *J. Chem. Phys.* **30**, 233 (1959).
[22] T. Azumi and S. P. McGlynn, *J. Chem. Phys.* **37**, 2413 (1962).
[23] V. G. Krishna and L. Goodman, *J. Chem. Phys.* **37**, 912 (1962).
[24] F. Dörr and H. Gropper, *Z. Elektrochem.* **67**, 193 (1963).
[25] M. A. El-Sayed, *Nature* **197**, 481 (1963).
[26] R. M. Hochstrasser and S. K. Lower, *J. Chem. Phys.* **40**, 1041 (1964).
[27] F. Dörr, H. Gropper, and N. Mika, *Z. Elektrochem.* **67**, 202 (1963).
[28] B. Ja. Sweschnikow and W. L. Jermolaev, *Ber. Akad. Wiss. Ud. SSR* [N.S.] **71**, 647 (1950).
[29] V. A. Pilipovich, *Opt. Spectry. (USSR) (English Transl.)* **10**, 104 (1961).

for fluorene, triphenylene, and coronene, but El-Sayed[25] investigated several aromatic hydrocarbons using the photoselection method and confirmed the observations made by other authors. The phosphorescence polarization of anthracene, pyrene, and other aromatics in benzophenone as a rigid solvent has been studied by Hochstrasser[26]; here too negative polarization has been observed. Heterocyclics with π–π^* phosphorescence such as quinoline, isoquinoline, 5,6- and 7,8-benzoquinoline, phenanthridine, and others likewise show negative polarization of phosphorescence according to the investigations of Dörr and Gropper.[19, 24, 27] Sweschnikow and Jermolaev[28] as well as Pilipovich[29] have established the same effect with dyestuffs such as auramine, fluorescein, and rhoduline orange.

As has been stated above, phosphorescence shows the same polarization as would be shown in an absorptive transition from the ground state to the singlet excited state perturbing the triplet state to cause the phosphorescence. The moment of this transition is therefore oriented perpendicular to the plane of the molecule. But such behavior is displayed only by transitions that involve σ electrons. It is therefore the surprising result of the measurements of the polarization of phosphorescence that the singlet excited state that mixes with and perturbs the triplet state is not a pure π-electron state. The nature of this σ term is not clear in any particular case. For the azahydrocarbons investigated by Dörr and Gropper[19] the obvious assumption is that a singlet n–π^* state is concerned. It is well known that such states are characterized by the promotion of one of the lone-pair electrons from an orbital of σ symmetry (in this case on the nitrogen) to an empty π^* orbital. Such n–π^* states with σ character are thus also to be assumed in azahydrocarbons even though the corresponding n–π^* transitions cannot be observed because of overlapping by other bands. For the hydrocarbons both the Rydberg and the ionic states (which also have σ character) have been considered to be the perturbing singlet terms.[30]

Among heterocyclics with n–π^* phosphorescence Krishna and Goodman[31] have observed for pyrazine and pyrimidine polarization in the plane of the molecule. The transition moment of the n–π^* phos-

[30] H. Gropper and F. Dörr, *Z. Elektrochem.* **67**, 46 (1963); M. A. El-Sayed, *Nature* **197**, 481 (1963); M. Mizushima and S. Koide, *J. Chem. Phys.* **20**, 765 (1952); E. Clementi, *J. Mol. Spectry.* **6**, 497 (1961).

[31] V. G. Krishna and L. Goodman, *J. Chem. Phys.* **36**, 2217 (1962).

phorescence is perpendicular to that of the n–π^* absorption. El-Sayed and Brewster[32] have generalized this result by showing that for heterocyclics with π–π^* phosphorescence the singlet term that mixes with the triplet state has σ character, in contrast to those with n–π^* phosphorescence, for which it has π character.

1.5. Triplet–Triplet Transitions

In principle every π-electronic configuration that describes an excited state can be resolved into a singlet and a triplet state. In practice with unsaturated organic compounds there are observed in addition to the lowest triplet excited state (which gives rise to the phosphorescence), higher triplet states that are of lower energy than the corresponding singlet states. Transitions between the lowest and these higher triplet states can occur as either radiating or radiationless processes. Since transitions between terms of the same multiplicity are thus involved, they take place with high probability, exactly like singlet–singlet transitions.

Because of the long lifetime of the lowest triplet state it is possible, by intense and continuous irradiation of a substance present in a rigid medium, to establish a relatively high stationary concentration of molecules that are in the triplet state. If an absorption spectrum is measured for a compound excited in this way, it is found that the singlet–singlet bands are lowered somewhat; beside them appear new bands, generally of longer wavelength, and these, after shutting off the irradiating light, slowly disappear again at a rate appropriate to the lifetime of the phosphorescence. These bands arise from absorptive transitions between the lowest triplet state and higher ones.

The first "absorption spectrum of a molecule in the triplet state" of this kind was measured by Lewis *et al.*[1] for fluorescein. Since then McClure[2] as well as Craig and Ross[3] have measured triplet–triplet absorption spectra of numerous polynuclear aromatic hydrocarbons, heterocyclics, and substitution products of aromatics. Naphthalene, anthracene, phenanthrene, chrysene, triphenylene, fluorene, 3,4-

[32] M. A. El-Sayed and R. G. Brewster, *J. Chem. Phys.* **39**, 1623 (1963).
[1] G. N. Lewis, D. Lipkin, and T. T. Magel, *J. Am. Chem. Soc.* **63**, 3005 (1941).
[2] D. S. McClure, *J. Chem. Phys.* **19**, 670 (1951).
[3] D. P. Craig and I. G. Ross, *J. Chem. Soc.* p. 1589 (1954).

benzpyrene, quinoline, isoquinoline, and halogenonaphthalenes have been investigated. The precise determination of the extinction coefficients of triplet–triplet absorption bands is difficult because of the uncertainty of the concentration of the "triplet molecule." Nevertheless the approximate values obtained (e.g., $\epsilon > 10,000$ for naphthalene,[3] $\epsilon > 50,000$ for anthracene,[3] though smaller for other compounds) make it clear that triplet–triplet transitions occur with the high probability to be expected.

If the lowest triplet state has a very short lifetime, as is the case with all compounds in liquid solution or in the gaseous state, then triplet–triplet spectra can be measured only if very high intensity of excitation is employed. An extremely elegant and very generally applicable method of measurement has been developed by Porter.[4] The substance, in liquid solution or the gaseous state, is excited by a light flash of very great intensity; in this way there is obtained for a short time a sufficiently high concentration of molecules in the lowest triplet state. Immediately after the exciting flash a second flash is passed through the solution perpendicular to the direction of excitation so that the light that passes through the solution falls onto a spectrograph. In this way triplet–triplet absorption spectra of numerous substances and under various external conditions could be measured. The method has also been used to follow very fast chemical reactions in which triplet molecules take part.

To identify the spectra obtained as of the triplet–triplet type, various criteria were employed. Thus the spectra obtained by flash spectroscopy were also measured in rigid solution where they decay with the same speed as the phosphorescence. Further, the energies of the higher triplet states of organic molecules can be calculated by the molecular orbital method and these theoretical values have proved very useful for identifying the measured spectra.

Triplet–triplet transitions can also take place without radiating (e.g., intramolecularly between a higher and the lowest triplet state). Of special interest is the phenomenon of a radiationless triplet energy transfer between two different sorts of molecule. In this an excited "donor" molecule in its lowest triplet state T_D and an "acceptor" molecule in its ground state N_A react with each other so that the acceptor goes over into its lowest triplet state T_A and the donor into the ground state N_D according to

$$T_D + N_A \rightarrow T_A + N_D$$

[4] G. Porter, *Proc. Chem. Soc.* p. 291 (1959).

The necessary prerequisite for a process of this type is that the triplet state of the donor have greater energy than that of the acceptor. A system that has been much investigated is that in which benzophenone is the donor and naphthalene the acceptor, their triplet energies being 72 and 61 kcal/mole, respectively.

In this and a series of similar systems the phenomenon of intermolecular triplet energy transfer has been discovered. If solutions of naphthalene in a rigid solvent at low temperature are irradiated with light of 366 mμ wavelength practically no phosphorescence is observed, since naphthalene only begins to absorb at about 325 mμ. Now Terenin and Ermolaev[5] found that under otherwise identical conditions but in the presence of benzophenone, which absorbs at 366 mμ and whose lowest triplet state lies higher than that of naphthalene, the characteristic green phosphorescence of the naphthalene develops intensely but that of benzophenone is almost completely suppressed. It is obvious that in the mechanism of this "sensitized phosphorescence" the transfer of triplet energy (from benzophenone to naphthalene) must play a part. Direct proof of this mechanism is found in the observation that the lifetime of the benzophenone phosphorescence is reduced in the presence of naphthalene.[5] The lifetime of the sensitized naphthalene phosphorescence is, in contrast, identical with that obtained by direct excitation.

In rigid solutions triplet energy transfer and sensitized phosphorescence are observable only at relatively high concentrations (greater than 10^{-2} mole/liter) since the intermolecular exchange is effective only over short distances. In liquid media and in gases triplet energy transfer is observed at greater dilutions[4,6] and the exchange between the molecules is here determined by diffusion.

It is worth noting that the quantum yield of the benzophenone-sensitized phosphorescence of naphthalene is greater than that resulting from direct excitation.[5] The especially high probability of intersystem crossing in benzophenone (no fluorescence) leaves its triplet state with a denser population than that naphthalene achieves by direct excitation. Transfer of the triplet energy from benzophenone to naphthalene follows, and this is almost complete in concentrated solution.

[5] A. Terenin and V. Ermolaev, *Trans. Faraday Soc.* **52**, 1042 (1956); *Proc. Acad. Sci. USSR* **85**, 547 (1952).

[6] G. Porter and F. Wilkinson, *Proc. Roy. Soc.* **A264**, 1 (1961).

Hochstrasser[7] has investigated the triplet energy transfer in mixed crystals of benzophenone and naphthalene. The phosphorescence arises exclusively from the naphthalene up to a concentration of 10^{-5} mole/per mole of the "guest" (naphthalene). Similar results were obtained when chrysene, pyrene, or anthracene replaced naphthalene as guest.[8]

An interesting study of this classical triplet energy transfer system, benzophenone–naphthalene, has been supplied by Hammond *et al.*[9] These authors investigated the luminescence properties of compounds of the type II. Triplet energy transfer can take place here *intramolecularly*

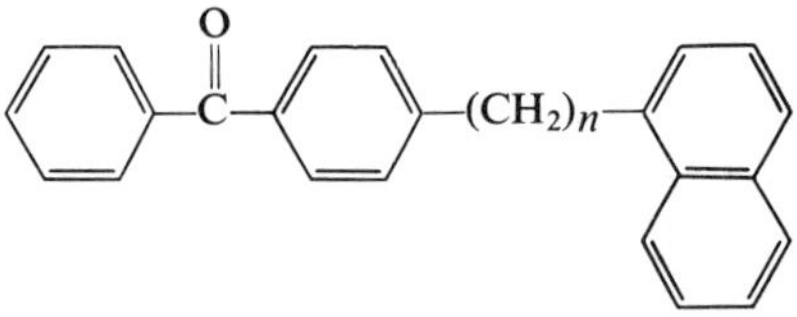

O
C
$(CH_2)_n$

II

between the chromophores benzophenone and naphthalene that are now separated by CH_2 groups. Irradiation with light of the wavelength 366 mμ, i.e., in the n–π^* band of the benzophenone, led exclusively to the sensitized naphthalene phosphorescence. It could be shown that the frequency of the process of triplet energy transfer at ca. 10^{10} times per second is greater than that of singlet energy transfers (see also Section 1.1). Like the corresponding intermolecular process, triplet energy transfer in II takes place by exchange interaction involving overlap of the electron clouds of the two unsaturated groups. Dexter[10] has presented the theory of this phenomenon as a modification of Förster's ideas.[11]

In principle then, sensitized phosphorescence ought always to be observable (at sufficiently high concentrations) if the following conditions are fulfilled:

$$T_A < T_D \text{ and } S_A > S_D$$

where T and S signify the lowest triplet and singlet states, respectively,

[7] R. M. Hochstrasser, *J. Chem. Phys.* **39**, 3153 (1963).
[8] R. M. Hochstrasser and S. K. Lower, *J. Chem. Phys.* **40**, 1041 (1964).
[9] A. A. Lamola, P. A. Leermakers, G. W. Byers, and G. S. Hammond, *J. Am. Chem. Soc.* **87**, 2322 (1965).
[10] D. L. Dexter, *J. Chem. Phys.* **21**, 836 (1953).
[11] Th. Förster, *Ann. Physik* [6] **2**, 55 (1948); *Discussions Faraday Soc.* **27**, 7 (1959).

and A and D the acceptor and donor. Of course both the donor and the acceptor may be aromatic hydrocarbons. A system that has been investigated very often is that of donor phenanthrene and acceptor naphthalene ($S_D = 29{,}000$, $T_D = 21{,}640$, $S_A = 31{,}400$, $T_A = 21{,}300\ \mathrm{cm}^{-1}$). Figure 7 shows the phosphorescence spectrum of a 1 : 1 mixture of 0.1 M

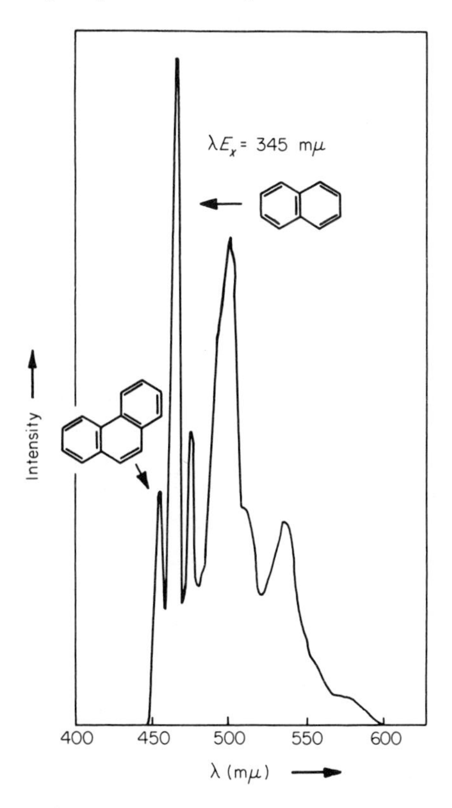

Fig. 7. Phosphorescence spectrum of a 1 : 1 mixture of 0.1 M solutions of phenanthrene and naphthalene in EPA on excitation with 345 mμ at 77°K. [According to M. Zander, *Erdoel Kohle* **19**, 278 (1966).]

solutions of phenanthrene and naphthalene in EPA at 77°K on irradiation in the longest-wavelength absorption band of phenanthrene (345 mμ). The phenanthrene phosphorescence appears only weakly in contrast with the intense sensitized phosphorescence of the naphthalene. The strong dependence of processes of triplet energy transfer on the concentration, which is universally displayed in rigid media, is shown in Table 7 for the phenanthrene–naphthalene system. Here are set out for

TABLE 7 Quenching of the Phosphorescence of Phenanthrene by Naphthalene[a]

Concentration (*M*)	Relative intensity of phosphorescence[b]	
	Phenanthrene	1:1 Phenanthrene–naphthalene
0.001	0.15	0.15
0.01	0.69	0.60
0.1	0.77	0.11

[a] M. Zander, *Erdoel Kohle* **19**, 278 (1966).
[b] EPA, 77°K.

various concentrations the relative intensities of phosphorescence of the phenanthrene in solution both as the sole solute and in the presence of an equal concentration of naphthalene. Although in the 0.1 *M* solution strong quenching of the phenanthrene phosphorescence by the naphthalene is found, it is only weak at 0.01 *M*, and at 0.001 *M* it is scarcely noticeable. An equally strong dependence on concentration has been found with many other systems.[12]

Hutchison *et al.*[13] have carried out an interesting ESR study on the phenanthrene–naphthalene system. The behavior of these compounds was investigated at low temperature in a single crystal of biphenyl. Irradiation in the longest-wavelength absorption band of phenanthrene produced unequivocally the triplet ESR spectrum of naphthalene.

From researches by Bhaumik and El-Sayed[14] and by Heller and Wassermann[15] it has been established that triplet energy transfer can take place also from an organic donor to a lanthanide element such as terbium or europium. When aromatic hydrocarbons and carbonyl

[12] M. Zander, *Erdoel Kohle* **19**, 278 (1966).
[13] R. W. Brandon, R. E. Gerkin, and C. A. Hutchison, Jr., *J. Chem. Phys.* **37**, 447 (1962).
[14] M. L. Bhaumik and M. A. El-Sayed, *J. Phys. Chem.* **69**, 275 (1965).
[15] A. Heller and E. Wassermann, *J. Chem. Phys.* **42**, 949 (1965).

compounds act as donors, the lanthanide acceptor exists either as a chelate[14] or as a salt of the tervalent ion.[15] The phosphorescence of the lanthanide ion is observed following irradiation in an absorption band of the organic donor.

1.6. Delayed Fluorescence

So far we have discussed the following processes which can start from the lowest triplet state (T_1) of a molecule: (1) phosphorescence (path 1, Fig. 8), (2) radiationless transition to the ground state (path 2), (3) triplet–triplet absorption (path 3), and (4) intermolecular triplet–triplet energy

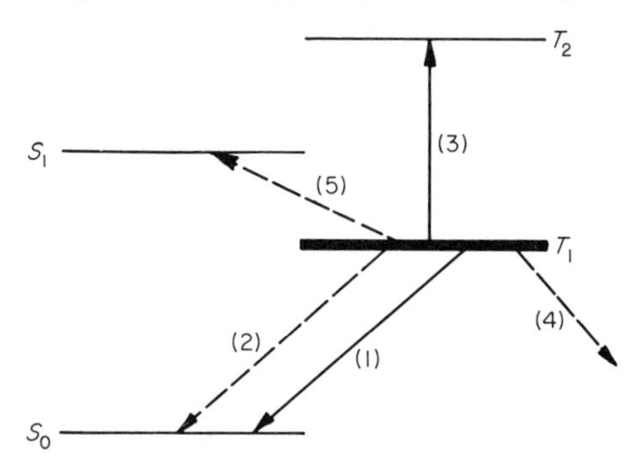

Fig. 8. Term scheme.

transfer (path 4). In principle, a further process seems to be possible, viz., a reversal of the intersystem crossing $S_1 \dashrightarrow T_1$ (path 5). A process of this kind (5), first suggested by Jablonski,[1] involves higher nuclear vibration terms of T_1 and requires the addition of external thermal energy. In contrast to light absorption it leads to delayed occupation of S_1. In systems in which this process takes place there ought to be two fluorescences, both starting from S_1 and so spectroscopically identical, viz., a "spontaneous" fluorescence excited by direct absorption of light and a "delayed" fluorescence excited thermally via T_1. It can be shown that the mean lifetime of this delayed fluorescence must be identical with that of the $T_1 \rightarrow S_0$ phosphorescence that appears simultaneously.[2]

[1] A. Jablonski, *Nature* **131**, 839 (1933); *Z. Physik* **94**, 38 (1935).

[2] Th. Förster, "Fluoreszenz organischer Verbindungen," p. 272ff. Vandenhoek & Ruprecht, Göttingen, 1951.

The activation energy of the thermal transition $T_1 \dashrightarrow S_1$ is given by the difference Δ between the energies of these terms. For compounds with large Δ such a process could only take place at relatively high temperatures, but at such temperatures quenching processes will usually be more probable than any others. For compounds with small Δ, however, the reversal of the $S_1 \dashrightarrow T_1$ transition seems to be always possible. Compounds with small Δ ought, therefore, to display spontaneous fluorescence, phosphorescence, and delayed fluorescence.

The phenomenon of thermally excitable delayed fluorescence has actually been known for a long time. In the older literature it is frequently described as "high-temperature phosphorescence." This expression is not very happily chosen and so the phrase "E-type delayed fluorescence," introduced by Parker and Hatchard,[3] is preferable. E stands for eosin, on which the phenomenon has been most thoroughly studied and which thus serves as the prototype of the compounds displaying it.

Parker and Hatchard[4] have investigated the luminescence properties of eosin in glycerol and ethanol over the temperature range +70° to −196°C. It shows, throughout this temperature range, a "spontaneous" fluorescence of short decay period. Measurements with the phosphoroscope reveal, in addition, both a delayed fluorescence with the same spectral distribution and a phosphorescence of longer wavelength. The lifetime of the delayed fluorescence and of the phosphorescence is of the order of a few milliseconds, is therefore relatively short, and is the reason that these luminescences can also be observed in liquid solutions up to relatively high temperatures. The radiant transitions proceed so rapidly that they can compete with the quenching processes that increase strongly in the liquid solutions. The excitation mechanism discussed at the beginning of this section for the E-type delayed fluorescence requires that its intensity increase with increasing temperature, while the intensity of the phosphorescence that simultaneously appears decreases. This is exactly what is observed with eosin. The thermal transition $T_1 \dashrightarrow S_1$ proceeds according to the first-order law and its activation energy can easily be determined if the ratio of the intensity of delayed fluorescence to that of phosphorescence is plotted logarithmically against the reciprocal of the absolute temperature. From the gradient of the resulting

[3] C. A. Parker and C. G. Hatchard, *Trans. Faraday Soc.* **59**, 284 (1963).
[4] C. A. Parker and C. G. Hatchard, *Trans. Faraday Soc.* **57**, 1894 (1961).

straight line Parker and Hatchard[4] estimated an activation energy of ca. 10 kcal/mole. This value agrees very accurately with the difference between the positions of the fluorescence and phosphorescence transitions, i.e., with the energy difference $S_1 - T_1$. It should be added that it is clear from the measurements made by Parker and Hatchard[4] that the frequencies of the two intersystem crossing processes $S_1 \dashrightarrow T_1$ and $T_1 \dashrightarrow S_1$ are of the same order of magnitude, namely 10^6–10^7 per second.

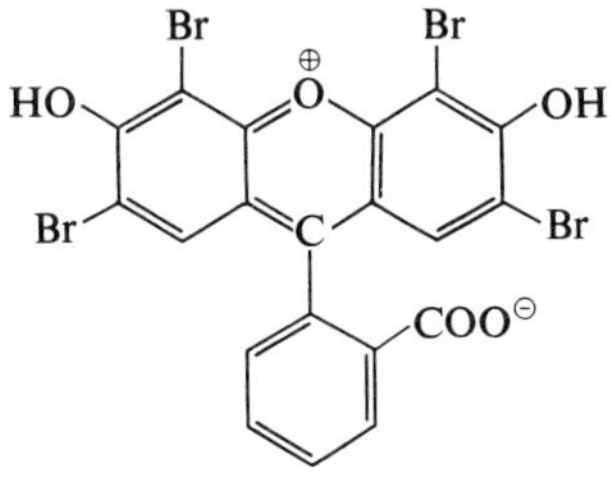

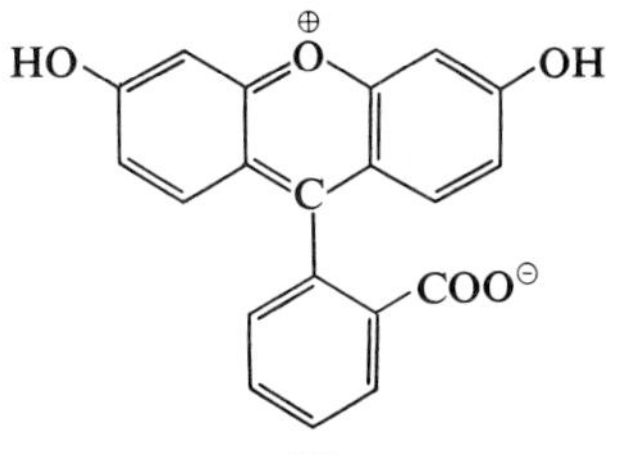

The majority of compounds for which E-type delayed fluorescence has been observed are found among the triphenylmethane and the acridine dyestuffs. Examples, in addition to eosin (III), are fluorescein (IV), acridine orange (V), and erythrosin (VI). For none of these compounds is the $S_1 - T_1$ energy difference greater than ca. 10 kcal/mole. For references to the voluminous older literature, the monographs of Förster[2] and Pringsheim[5] should be consulted.

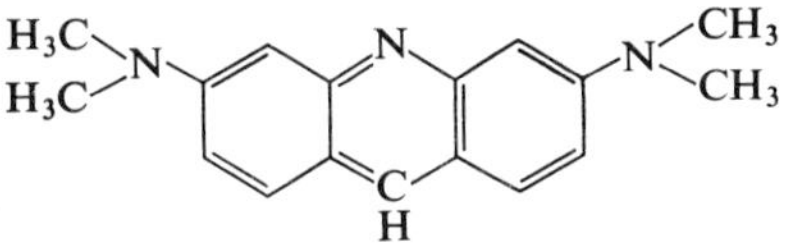

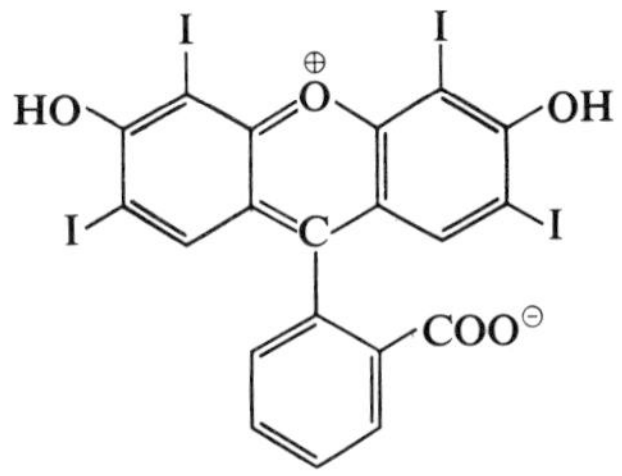

The unsubstituted hydrocarbons do not conform to this pattern, however. For the majority of aromatics the energy difference $S_1 - T_1$

[5] P. Pringsheim, "Fluorescence and Phosphorescence," p. 285ff. Wiley (Interscience), New York, 1963.

lies between 20 and 30 kcal/mole. In these compounds the transition $T_1 \dashrightarrow S_1$ can first take place at temperatures at which quenching processes are more probable than any others. Thus E-type delayed fluorescence is not characteristic of aromatic hydrocarbons.[6]

So far only one exception is known. For coronene the difference $S_1 - T_1$ comes to about 12 kcal/mole. As Zander has shown,[7] rigid solutions of coronene in perhydrocoronene ($C_{24}H_{36}$) at -196°C give almost exclusively the familiar T_1–S_0 phosphorescence. At room temperature, beside the phosphorescence, the very faint delayed fluorescence of the coronene can just be detected; at 100°C it has become very intense and at 170°C it is the only luminescence observable. Although a rigorous proof is still lacking—it might be secured by measuring the activation energy of the delayed fluorescence and determining the lifetimes accurately—it is possible, because of its strong dependence on the temperature, to assume that here the E-type delayed fluorescence really is excited in the way discussed initially, i.e., by Jablonski's mechanism. The remarkable insensitivity of the triplet state to thermal quenching processes seems to be a special property of the coronene–perhydrocoronene system and must be connected with its crystal structure.

A type of delayed fluorescence different from that discussed so far was first observed by Kautsky and Müller.[8] Adsorbates of tryptaflavin on silica gel at low temperatures display only spontaneous fluorescence and a yellow $T_1 \rightarrow S_0$ phosphorescence. If small amounts of oxygen are added to the adsorbates excited at low temperatures, the phosphorescence is quenched and a delayed fluorescence appears without further excitation. This has the same spectral distribution as the spontaneous fluorescence. At the low temperature at which delayed fluorescence is here observed, it is impossible for a thermal transition $T_1 \dashrightarrow S_1$ to take place to any significant extent.* The energy necessary for the occupation of S_1 from T_1 must therefore be obtained by some other mechanism. Kautsky[8] has proposed one in which *two* molecules that are in the triplet state react with each other by exchange so that one

* But see l.c.[2]

[6] H. Kautsky and H. Merkel, *Naturwissenschaften* **27**, 195 (1939).

[7] M. Zander, *Naturwissenschaften* **47**, 443 (1960).

[8] H. Kautsky and G. O. Müller, *Naturwissenschaften* **29**, 150 (1941); *Z. Naturforsch.* **2a**, 167 (1947).

molecule, by taking up the triplet energy of the other, goes over into the first singlet excited state, the second going to the ground state; the oxygen therefore serves only as a means of taking up the excess energy $2T_1 - S_1$. A reaction of this sort is formulated as

$$T_1 + T_1 \rightarrow S_1 + S_0 \tag{3}$$

This is really nothing more than an energy balance, but experimental proofs that a reaction of this kind is possible were first provided (for another system) by Parker and Hatchard.[9]

These authors observed that solutions of anthracene or phenanthrene in carefully deoxygenated ethanol at room temperature produce, as well as the spontaneous fluorescence, a delayed fluorescence which can be detected with a phosphoroscope. Its lifetime is of the order of magnitude of several milliseconds. In spectral distribution it agrees perfectly with the spontaneous fluorescence. The excitation mechanism of E-type delayed fluorescence breaks down because of the great difference between S_1 and T_1 for the compounds under consideration and because of the relatively low temperature at which the delayed fluorescence was observed. One decisive observation points to the Kautsky mechanism: the intensity of the delayed fluorescence is proportional to the *square* of the intensity of the absorbed exciting light. This is in accord only with a mechanism by which, through energy transfer, *two* excited molecules produce *one* molecule capable of fluorescence. The observation asserts nothing with respect to whether there are two singlet-excited molecules involved, two triplets, or one singlet and one triplet.

Nevertheless the Kautsky mechanism (interaction of two triplets) is the most adequate. There are at least two experimental criteria by which it can be distinguished from the other two possibilities (reaction between two singlets and between one singlet and one triplet). In systems in which phosphorescence can be observed as well as delayed fluorescence of the type now under discussion, the phosphorescent lifetime ought (as a simple application of kinetic theory easily shows) to be twice as long as that of the delayed fluorescence; moreover, the intensity of the delayed fluorescence must be proportional to the *square* of that of the phosphorescence.

Stevens and Walker[10] have provided the experimental proof of the

[9] C. A. Parker and C. G. Hatchard, *Proc. Roy. Soc.* **A269**, 574 (1962).
[10] B. Stevens and M. S. Walker, *Proc. Chem. Soc.* p. 181 (1963).

validity of the relation

$$\tau_{DF} = \tfrac{1}{2}\tau_P \tag{4}$$

Solutions of pyrene in liquid paraffin show spontaneous fluorescence, phosphorescence, and delayed fluorescence even at temperatures at which the E-type mechanism is excluded. Measurement of the decay times of phosphorescence and of delayed fluorescence showed that they obeyed the above relation.

Finally McGlynn and Azumi[11] were able to demonstrate conclusively for yet other systems that the intensity of the kind of delayed fluorescence discussed here is proportional to the *square* of the intensity of phosphorescence.

It has, therefore, been experimentally established that the excitation mechanism first postulated by Kautsky[8] in fact does operate. The exchange reaction between two molecules in the lowest triplet state by which one goes into a singlet excited state and the other into the ground state is generally described nowadays as "triplet–triplet annihilation." The delayed fluorescence that is associated with this has been called "P-type delayed fluorescence" by Parker and Hatchard,[3] the P standing for pyrene, which, as indicated above, shows it and has been very thoroughly investigated in connection with it.

McGlynn and Azumi[11] and others have shown that the course of the triplet–triplet annihilation process described by Eq. (3) should not be regarded as one of simple emission–absorption in which one molecule in the triplet state takes up the phosphorescent radiation of the second after the manner of a triplet–triplet absorption and then passes from a higher triplet state by intersystem crossing into the first excited singlet state. In Eq. (3) the process is much more like a resonance interaction which can only be properly described quantum mechanically.

P-Type delayed fluorescence is a very widespread phenomenon, and it is shown in the vapor state by numerous aromatic hydrocarbons such as phenanthrene,[12,13] anthracene,[13,14] perylene,[13,14] and pyrene.[13,14]

The delayed fluorescence of liquid solutions of aromatic hydrocarbons has already been referred to.[9,15] In both liquid solutions and vapors the

[11] T. Azumi and S. P. McGlynn, *J. Chem. Phys.* **38**, 2773 (1963); **39**, 1186 (1963).
[12] P. P. Dikun, *Zh. Eksperim. i. Teor. Fiz.* **20**, 193 (1950).
[13] R. Williams, *J. Chem. Phys.* **28**, 577 (1958).
[14] B. Stevens, M. S. Walker, and E. Hutton, *Proc. Chem. Soc.* p. 62 (1963).
[15] B. Stevens and E. Hutton, *Nature* **186**, 1045 (1960); C. A. Parker and C. G. Hatchard, *Proc. Chem. Soc.* p. 147 (1962).

process of triplet–triplet annihilation is controlled by diffusion so that the phenomenon is also observed in very dilute gases and solutions. In rigid glasses, e.g., EPA at low temperatures, McGlynn and Azumi[11] found that beside the phosphorescence that is very intense under these conditions a delayed fluorescence also appears if the concentration is relatively high ($> 0.01\ M$). Delayed fluorescence in EPA glass has been observed for naphthalene, naphthalene-d_8, phenanthrene, triphenylene, and biphenyl.

One of the first observations of delayed fluorescence with aromatic hydrocarbons was made on "pure" crystals of naphthalene and phenanthrene at low temperatures.[16] Today it ought to be regarded as certain, at least for naphthalene, that one is not concerned here with the delayed fluorescence of the pure compound but rather with that of a small amount of impurity (2-methylnaphthalene).[17] Also the delayed fluorescence of the guest molecule comes about by triplet–triplet annihilation, the theory of which has been developed for mixed crystals particularly by Robinson *et al.*[17] In the investigation of the delayed fluorescence of aromatic mixed crystals, for example of the type of chrysene (guest) and phenanthrene (host), it has been established that the intensity of the delayed fluorescence depends in an interesting way on the temperature.[18] This can be attributed to a radiationless transition from the lowest triplet state of the guest to the lowest triplet state of the host. The latter undertakes the transfer of the energy of one guest triplet to another and in this way takes part in the triplet–triplet annihilation and in the establishment of the delayed fluorescence. Other processes also play a part.

A phenomenon that is also of analytical significance and therefore of special interest for the field of this monograph is the sensitized P-type delayed fluorescence.[9, 19] At room temperature 0.001 M solutions of

[16] H. Sponer, Y. Kanda, and L. A. Blackwell, *J. Chem. Phys.* **29**, 721 (1958); H. Sponer, *in* "Luminescence of Organic and Inorganic Materials" (H. Kallmann and G. M. Spruch, eds.), p. 143. Wiley, New York, 1962; N. W. Blake and D. S. McClure, *J. Chem. Phys.* **29**, 722 (1958).

[17] H. Sternlicht, G. C. Nieman, and G. W. Robinson, *J. Chem. Phys.* **38**, 1326 (1963).

[18] M. Zander, *Z. Elektrochem.* **68**, 301 (1964); L. Azarraga, T. N. Misra, and S. P. McGlynn, *J. Chem. Phys.* **42**, 3720 (1965); N. Hirota and C. A. Hutchison, *ibid.* p. 2869; F. Dupuy, R. Lochet, and A. Rousset, *Compt. Rend.* **258**, 4223 (1964).

[19] C. A. Parker, *Proc. Roy. Soc.* **A276**, 125 (1963); C. A. Parker and C. G. Hatchard, *Proc. Chem. Soc.* p. 386 (1962).

phenanthrene to which very tiny quantities of anthracene have been added show only the *spontaneous* fluorescence of the phenanthrene. If the luminescence of these solutions is examined with the phosphoroscope it is surprising to find only the *delayed* fluorescence of the anthracene. Clearly energy transfer takes place from the triplet state of the phenanthrene to the lower-lying triplet state of the anthracene (see Section 1.5). Thus anthracene molecules in the triplet state give rise (in the sense of the triplet–triplet annihilation mechanism) to delayed fluorescence. In this way it is possible to sensitize the delayed fluorescence of compounds selectively. The analytical application of this phenomenon[20] will be looked into later (Section 3.8).

Finally it should be mentioned that, in addition to E- and P-type delayed fluorescence, there is a third type, which has been investigated very thoroughly by Stevens and Walker[21] for perylene. These authors found that rigid solutions of this compound in paraffin at 77°K showed delayed as well as spontaneous fluorescence. McGlynn and Azumi[11] had also used rigid solutions in their researches and it had to be assumed that the delayed fluorescence observed with perylene was likewise of P type. In fact McGlynn's and Stevens's systems and results differ in two important respects: the delayed fluorescence of perylene was observable even in dilute solutions (10^{-6} *M*) and could only be excited by light of wavelength less than 280 mμ. But in this spectral region, as Porter and Windsor have shown,[22] perylene ionizes from the triplet state. The obvious suggestion is, therefore, that the delayed fluorescence observed by Stevens and Walker is a "recombination glow"; the perylene molecules become photoionized and the separated electrons are trapped in the solvent matrix until they are able to be reunited with the perylene cation by emission of a delayed fluorescence. Similar mechanisms have been assumed for the delayed fluorescence of acriflavin[23] and polymethinecyanine dyes[24] in rigid glasses at low temperatures.

[20] C. A. Parker, C. G. Hatchard, and T. A. Joyce, *Analyst* **90**, 1 (1965).

[21] B. Stevens and M. S. Walker, *Chem. Commun.* p. 8 (1965); but see C. A. Parker and T. A. Joyce, *J. Chem. Soc.* p. 821 (1966).

[22] G. Porter and M. W. Windsor, *Discussions Faraday Soc.* **17**, 178 (1954).

[23] E. C. Lim and G. W. Swenson, *J. Chem. Phys.* **36**, 118 (1962).

[24] J. Kern, F. Dörr, and G. Scheibe, *Z. Elektrochem.* **66**, 462 (1962).

APPENDIX: REFERENCES TO REVIEW LITERATURE ON TRIPLET SPECTROSCOPY

Since the appearance of the fundamental work of Lewis and Kasha [*J. Am. Chem. Soc. 66*, 2100 (1944)] on the connection between phosphorescence and the triplet states of unsaturated organic molecules, the

TABLE 8 Review Articles on Triplet Spectroscopy

Phosphorescence

1. M. Kasha, *Chem. Rev.* **41**, 401 (1947).
2. M. Kasha and S. P. McGlynn, *Ann. Rev. Phys. Chem.* **7**, 403 (1956).
3. C. Reid, *Quart. Rev.* (*London*) **12**, 205 (1958).
4. M. Kasha, *Radiation Res.* Suppl. 2, 243 (1960).
5. S. K. Lower and M. A. El-Sayed, *Chem. Rev.* **66**, 199 (1966).

Photochemistry and the Triplet State

1. G. Porter, *Proc. Chem. Soc.* p. 291 (1959).
2. R. Hochstrasser and G. B. Porter, *Quart. Rev.* (*London*) **14**, 146 (1960).

Luminescence in Crystals

1. R. M. Hochstrasser, *Rev. Mod. Phys.* **34**, 531 (1962).
2. H. Sternlicht, G. C. Nieman, and G. W. Robinson, *J. Chem. Phys.* **38**, 1326 (1963).

Energy Transfer

1. Th. Förster, *Radiation Res.* Suppl. 2, 326 (1960).
2. M. Kasha, *Radiation Res.* **20**, 55 (1963).

n–π^* Transitions

1. J. W. Sidman, *Chem. Rev.* **58**, 689 (1958).

Luminescence of Molecular Compounds

1. S. P. McGlynn, *Chem. Rev.* **58**, 1113 (1958).

field of triplet spectroscopy has developed with extraordinary rapidity and has spread, fanwise, into many subdivisions. Thus not only spectroscopists have taken part in its expansion, but also photochemists, solid-state physicists, biochemists, and others, and it is not surprising, therefore, that the relevant literature has grown like an avalanche.

In the foregoing pages the theoretical and experimental bases of triplet spectroscopy have been discussed only insofar as seems necessary for the analytical chemist who wishes to appraise, develop, and apply the methods of phosphorimetric analysis. Those who have special interests must refer to the quite voluminous literature. However, to facilitate the approach to original publications a list of review articles on various aspects of triplet spectroscopy is given in Table 8.

CHAPTER 2 / PHOSPHORESCENCE

PROPERTIES OF INDIVIDUAL COMPOUNDS

The phosphorescence properties of individual organic compounds are described in the following sections according to their chemical classes, and the discussion of the experimental results has been placed prominently in the foreground so that relationships between constitution and phosphorescence behavior may be established as far as possible. However, in this summary it has not been possible to include all the compounds that have been investigated, nor even all the classes of compounds. In particular the whole broad field of the organic dyestuffs has been excluded; for this, reference should be made to the monographs of Förster[1] and Pringsheim.[2] Special care has been taken, however, to include those classes and individual substances that seem to be of greater interest because of the application of their phosphorescence in analysis.

2.1. Aromatic Hydrocarbons and Their Homologs

Benzene, naphthalene, and the linearly annellated hydrocarbons anthracene (VII), tetracene (VIII), pentacene (IX), etc. (the *acenes*), are of special significance for the understanding of the spectroscopic properties of aromatic hydrocarbons.

The phosphorescence spectrum of benzene has been studied not only by Lewis and Kasha[3] but, with special thoroughness, by Shull[4] and

[1] Th. Förster, "Fluoreszenz organischer Verbindungen," p. 261ff. Vandenhoeck & Ruprecht, Göttingen, 1951.

[2] P. Pringsheim, "Fluorescence and Phosphorescence," p. 285ff. Wiley (Interscience), New York, 1963.

[3] G. N. Lewis and M. Kasha, *J. Am. Chem. Soc.* **66**, 2100 (1944).

[4] H. J. Shull, *J. Chem. Phys.* **17**, 295 (1949).

Sveshnikov[5] using EPA at 77°K and ethanol at 90°K respectively. The vibrational analysis that these authors carried out independently of each other revealed that the phosphorescence transition of benzene is symmetry forbidden (see Section 1.3). The question of the assignment of the *S–T* transition has been the subject of much experimental and theoretical investigation. Kearns[6] has presented a particularly painstaking and detailed discussion, including consideration of the previous

VII VIII

IX

literature, and confirmed the assignment already given by Shull and by Dikun and Sveshnikov, ${}^3B_{1u} \rightarrow {}^1A_{1g}$. (See Sandorfy[7] for the group theoretical description of the electronic states of organic molecules.)

The phosphorescence spectra of benzene both in EPA and in a cyclohexane matrix at 77°K as measured by Sponer *et al.*[8] are reproduced in Fig. 9. The low intensity of the symmetry-forbidden 0,0 band is quite striking in both media and so too is the marked resolution of the spectrum in the cyclohexane matrix; this develops still further until, at 4°K, 124 separate bands can be observed (see Section 1.3 in this connection).

Measurements of the singlet–triplet absorption spectrum of benzene (as a vapor and as a liquid) have been carried out by several authors.[9, 10] The oxygen method developed by Evans (see Section 1.2) has been

[5] P. P. Dikun and B. Ya. Sveshnikov, *Dokl. Akad. Nauk SSSR* **65**, 637 (1949); *Zh. Eksperim. i Teor. Fiz.* **19**, 1000 (1949).

[6] D. R. Kearns, *J. Chem. Phys.* **36**, 1608 (1962).

[7] C. Sandorfy, "Die Elektronenspektren in der theoretischen Chemie," p. 41ff. Verlag Chemie, Weinheim, 1961.

[8] H. Sponer, Y. Kanda, and L. A. Blackwell, *Spectrochim. Acta* **16**, 1135 (1960).

[9] D. F. Evans, *J. Chem. Soc.* p. 3885 (1957).

[10] D. F. Evans, *J. Chem. Soc.* p. 1351 (1957); G. N. Lewis and M. Kasha, *J. Am. Chem. Soc.* **67**, 994 (1945); A. Pitts, *J. Chem. Phys.* **18**, 1416 (1950).

applied to this, sometimes intentionally and sometimes unintentionally (by earlier workers).

According to various writers[4, 5] the 0,0 band of the *S–T* transition of benzene in emission lies between 29,470 and 29,515 cm^{-1} (in a cyclohexane matrix[8] it is at 29,501 cm^{-1}) and in absorption at 29,510 cm^{-1} in the vapor[9] and between 29,400 and 29,440 cm^{-1} in the liquid.[10]

The *S–T* transition of naphthalene is permitted by symmetry and is found at 21,246 cm^{-1} in emission[11] (light petroleum, 77°K) and at 21,180 cm^{-1} in absorption using the oxygen method.[12] Vibrational

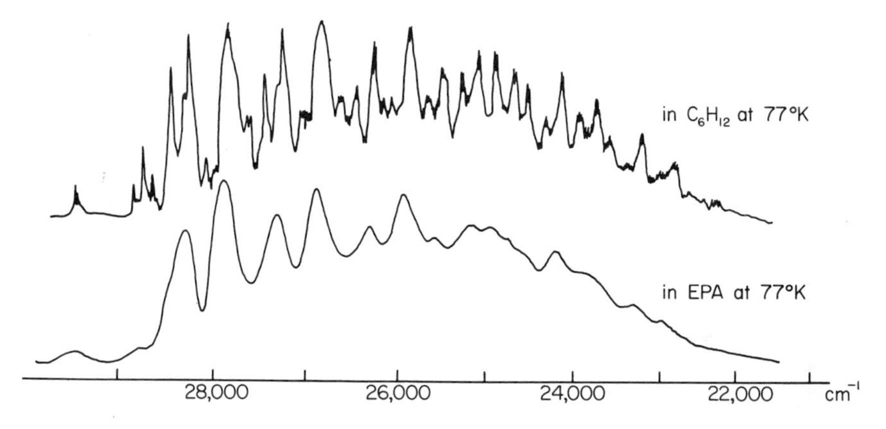

Fig. 9. Phosphorescence spectra of benzene in cyclohexane and EPA at 77°K [according to H. Sponer, Y. Kanda, and L. A. Blackwell, *Spectrochim. Acta* **16**, 1135 (1960)].

analyses of the phosphorescence spectrum have been given by several workers.[11, 13]

Because of its low intensity and its long wavelength the phosphorescence spectrum of anthracene (VII) was very difficult to determine, but was first measured by Lewis and Kasha.[3] Inconsistent results obtained by Reid[14] turned out to be incorrect and the old measurement

[11] J. Ferguson, T. Iredale, and J. A. Taylor, *J. Chem. Soc.* p. 3160 (1954).
[12] D. F. Evans, *J. Chem. Soc.* p. 1351 (1957).
[13] J. Czekalla, G. Briegleb, W. Herre, and H. J. Vahlsensieck, *Z. Elektrochem.* **63**, 715 (1959); J. W. Sidman, *J. Chem. Phys.* **25**, 229 (1956).
[14] C. Reid, *J. Chem. Phys.* **20**, 1212 (1952); *J. Am. Chem. Soc.* **76**, 3264 (1954).

by Lewis and Kasha[3] was established by Kasha *et al.*[15] and supplemented by measurement of the S–T absorption spectrum of anthracene. In emission its S–T transition lies at 14,927 cm^{-1} (EPA, 77°K) and in absorption at 14,850 cm^{-1} (CS_2, room temperature).

The S–T transition of tetracene (VIII) has so far only been measured in absorption[15] and is found—in agreement with quantum mechanical predictions[16]—at 10,250 cm^{-1}. It has not so far proved possible to observe the corresponding phosphorescence because, as would be expected, its intensity is very low. A tetracene phosphorescence reported by Reid[17] to lie at 19,600 cm^{-1} (5100 Å) that led to far-reaching conclusions was shown by Clar and Zander[18] and later by Hammond *et al.*[19] to be caused by a very small trace of tetracene-5,12-quinone present as an impurity.

For the higher acenes—pentacene (IX) and so on—attempts to evaluate the singlet–triplet transitions have succeeded neither in absorption nor in emission. However, the triplet–triplet absorptions of pentacene (and of the earlier members of the acene series) have been measured.[20]

The phosphorescence transition is displaced strongly and progressively toward longer wavelengths in the series benzene, naphthalene, anthracene, tetracene, i.e., with increasing linear annellation. If the position of the S–T transitions of these compounds is plotted against their ionization energy, there is obtained the close approximation to a linear relationship, shown in Fig. 10.[21] The equation of the straight line shown is

$$\tilde{\nu}_{S\text{–}T} = I - 46{,}000 \quad (\text{cm}^{-1}) \tag{5}$$

and implies that the positions of the S–T transitions of the acenes depend essentially on the fact that the energy of the ground state decreases as

[15] S. P. McGlynn, M. R. Padhye, and M. Kasha, *J. Chem. Phys.* **23**, 593 (1955); **24**, 588 (1956); S. P. McGlynn, T. Azumi, and M. Kasha, *ibid.* **40**, 507 (1964).
[16] R. Pariser, *J. Chem. Phys.* **24**, 250 (1956); G. G. Hall, *Proc. Roy. Soc.* **A213**, 113 (1952).
[17] C. Reid, *J. Chem. Phys.* **20**, 1214 (1952).
[18] E. Clar and M. Zander, *Chem. Ber.* **89**, 749 (1956).
[19] A. A. Lamola, W. G. Herkstroeter, J. C. Dalton, and G. S. Hammond, *J. Chem. Phys.* **42**, 1715 (1965); E. Clar and M. Zander, *ibid.* **43**, 3422 (1965).
[20] G. Porter, *Proc. Chem. Soc.* p. 291 (1959).
[21] M. Zander, Dissertation, Münster (1956); *Angew. Chem.* **72**, 513 (1960).

the annellation increases, whereas the energy of the lowest triplet state is almost entirely independent of the size of the molecule. If the relationship given by Eq. (5) or by Fig. 10 is extrapolated to the higher acenes, it is found that with ever increasing annellation the *S–T* transition moves to ever longer wavelengths until finally, at about nonacene, the still hypothetical hydrocarbon with nine linearly annellated benzene rings, the lowest triplet state should coincide with the ground state. Hence for these very highly annellated acenes we must be dealing with free radicals.

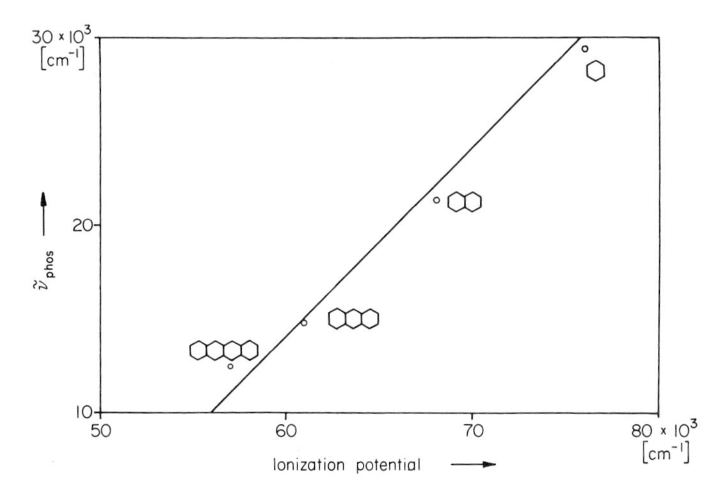

Fig. 10. Relationship between the positions of the phosphorescence transitions and the ionization energy for aromatic hydrocarbons.

For hexacene (X) and heptacene (XI), which are both known, the energy of the *S–T* transition estimated from Eq. (5) already falls in the region of *thermal* energy. In principle, therefore, it ought to be possible, if the considerable experimental difficulties could be overcome, to excite the lowest triplet states of these compounds thermally.

Among the phenes (the angularly annellated hydrocarbons phenanthrene (XII), tetraphene (XIII), pentaphene (XIV), etc.) phenanthrene[18, 22] has been investigated very thoroughly. The 0,0 band of the spectrum of its intense green phosphorescence lies at 21,600 cm^{-1} in

[22] M. Kasha, *Chem. Rev.* **41**, 401 (1947); D. S. McClure, *J. Chem. Phys.* **17**, 905 (1949); J. Czekalla and K. J. Mager, *Z. Elektrochem.* **66**, 65 (1962).

EPA at 77°K (see Fig. 5), and vibrational analyses have been given by several authors.[23] Evans[12] has measured the S–T absorption spectrum (0,0 band at 21,600 cm^{-1}).

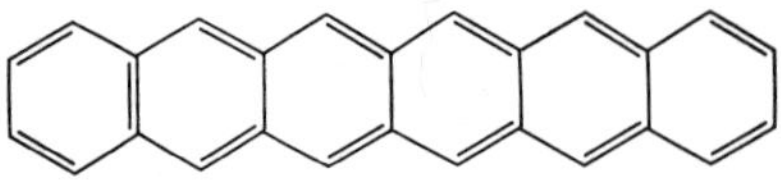

X

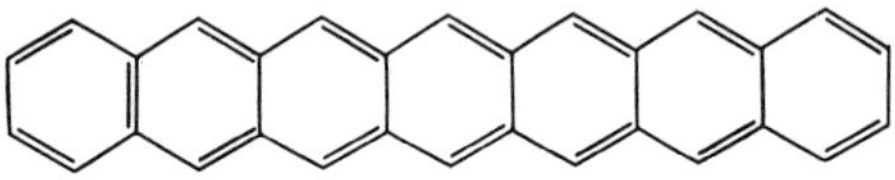

XI

With the phenes too the phosphorescence transition is displaced toward longer wavelengths with increasing linear annellation of the longer arm. Thus the 0,0 bands of the phosphorescence spectra of tetraphene and pentaphene (each with three rings in the longer arm) are found[18] at 16,520 and 16,930 cm^{-1}, respectively, whereas that of phenanthrene (two rings maximum) is at 21,600 cm^{-1}.

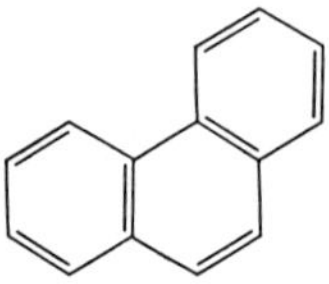

XII 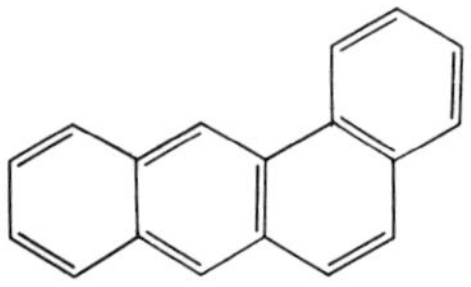 XIII 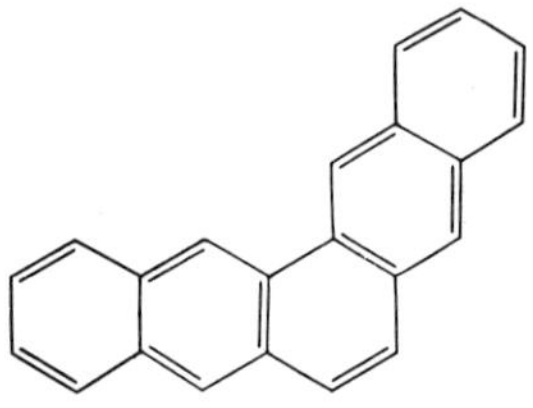 XIV

The classification of the lowest triplet states of the aromatic hydrocarbons is of special importance (in this connection, see Section 1.3). In Fig. 11 the positions of the α, β, and para absorption bands (^1L_b, ^1B_b, ^1L_a), with those of the S–T transitions of the acenes and phenes, have been plotted as functions of the number of rings. It can be seen that there is a convincing correlation between S–T bands and para bands, although with the α and β bands no relationship can be established. The correlation between S–T bands and para bands has been confirmed by a very

[23] Y. Kanda and R. Shimada, *Spectrochim. Acta* **15**, 211 (1959); J. Czekalla, G. Briegleb, W. Herre, and H. J. Vahlsensieck, *Z. Elektrochem.* **63**, 715 (1959).

large amount of experimental evidence.[18] Figure 12 shows, for further aromatic hydrocarbons of various structural patterns, the parallel dependence of both transitions on the constitutions of the molecules. From the relationship presented in Figs. 11 and 12 it must be concluded that the lowest triplet states of aromatic hydrocarbons must have the same electronic configuration as the para singlet excited state, i.e., it must be classified as ^3L_a.[24] Kearns[6] has unequivocally confirmed this.

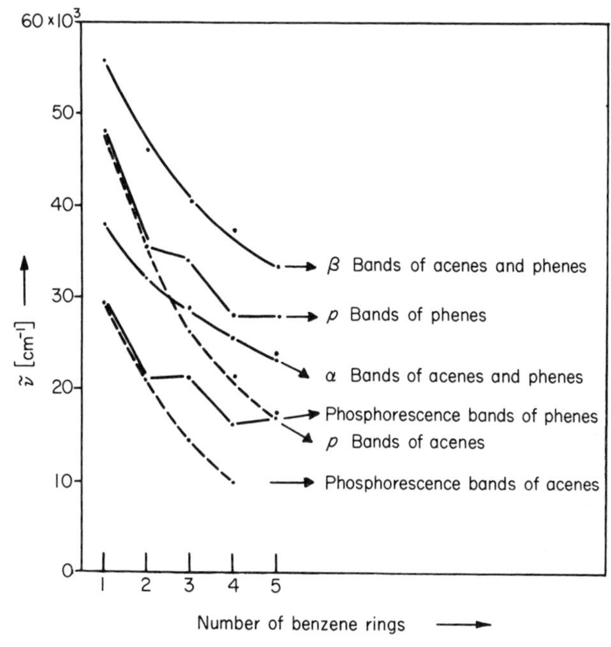

Fig. 11. Dependence of the position of the *S–S* and *S–T* transitions of acenes and phenes on the number of annellated benzene rings.

As a criterion of classification we have made use of the fact that the singlet and triplet states that belong to the same electronic configuration show the same dependence on constitution, i.e., that both terms are displaced parallel to each other in passing from one member of a series to another, or, expressing it more simply, that the magnitude of the singlet–triplet splitting within a given class of substances, e.g., aromatic hydrocarbons, is independent of the size and structure of the molecule.

[24] I. R. Platt, *J. Chem. Phys.* **17**, 484 (1949); H. B. Klevens and I. R. Platt, *ibid.* p. 470.

This is, as follows from Figs. 11 and 12, quite a good approximation, but closer inspection shows that it is not exactly true. In reality it is observed that the singlet–triplet splitting in aromatic hydrocarbons does depend on the molecular size. In Fig. 13 the differences (in cm^{-1}) between

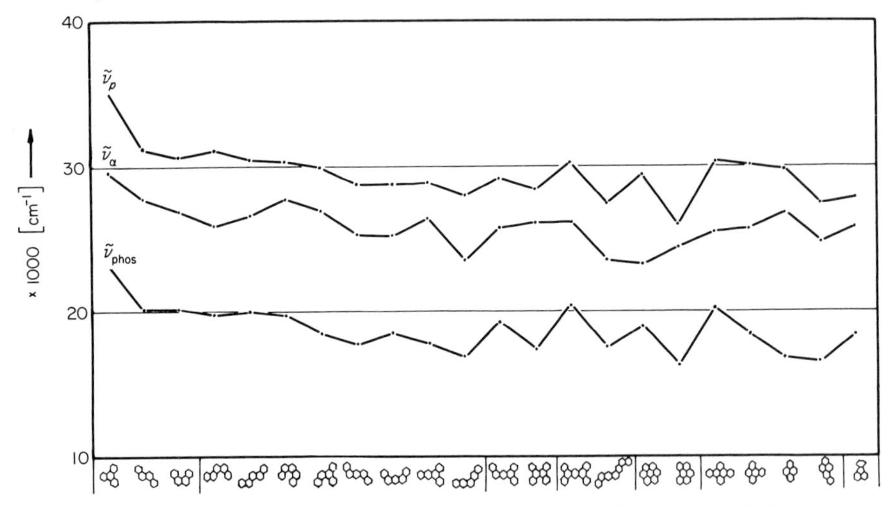

Fig. 12. Positions of the phosphorescence transitions and of the α and the para absorptions of polycyclic aromatic hydrocarbons.

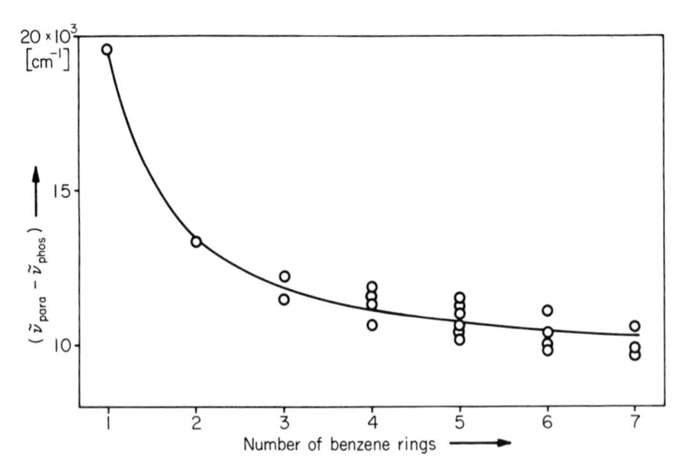

Fig. 13. Relationship between *S–T* splitting and molecular size for aromatic hydrocarbons.

the 0,0 bands of para absorption and phosphorescence for numerous aromatic hydrocarbons of various structural types have been plotted against the number of benzene rings present in the molecule, since this can be regarded as a measure of the size of the molecule.[21,25] It is clear that this difference, the singlet–triplet splitting, falls smoothly as the dimensions of the molecule increase, and becomes approximately constant (at ca. 10,000 cm^{-1}) only for hydrocarbons with more than five rings. The relationship represented by Fig. 13 can be used to predict the approximate molecular weight from knowledge of the positions of the para and *S–T* bands of an unknown hydrocarbon, or, more important, to predict from the absorption spectrum of a known hydrocarbon the approximate position of its phosphorescence spectrum. This is often useful in planning spectrophosphorimetric analyses.

TABLE 9 Phosphorescence of Polycyclic Aromatic Hydrocarbons[a]

Compound[b]	0,0 Band (cm^{-1})	Reference[c]	Lifetime[d] (sec.)
	a. Acenes and Phenes		
Benzene	29,500	3, 4, 5, 8	7.0
Naphthalene	21,246	3, 11, 13	2.6
Anthracene (VII)	14,927	3, 15	0.09
Tetracene (VIII)	10,250 (absorption)	15	—
Phenanthrene (XII)	21,600	18, 22	3.3
Tetraphene (XIII)	16,520	18	0.3
Pentaphene (XIV)	16,930	18	—
	b. kata-Annellated, Angular Hydrocarbons		
Chrysene	20,040	3, 18	2.5
3,4-Benzphenanthrene (XLV)	20,020	18, 55, 56	3.5
1,2-Benzchrysene	18,650	18	—
Picene	20,080	18	2.5[e]
5,6-Benzchrysene	19,760	18	—
3,4:5,6-Dibenzphenanthrene	19,800	18	—
1,2:7,8-Dibenzchrysene	17,270	18	—
1,2:5,6-Dibenzanthracene	18,260	18	1.5
1,2:3,4-Dibenzanthracene	17,770	18	—
1,2:7,8-Dibenzanthracene	18,500	18	—

[25] M. Zander, *Angew. Chem. Intern. Ed. Engl.* **4**, 930 (1965).

Compound[b]	0,0 Band (cm^{-1})	Reference[c]	Lifetime[d] (sec.)
c. Hydrocarbons of the Pyrene and Perylene Series			
Pyrene (XVI)	16,850	14, 27, 31	0.2
1,2-Benzpyrene (XVII)	18,510	18, 31	2.0
3,4-Benzpyrene (XVIII)	14,670	28	—
1,2:6,7-Dibenzpyrene (XIX)	20,360	18	7.5
1,2:4,5-Dibenzpyrene (XX)	16,360	29	—
1,12-Benzperylene (XXIII)	16,180	18	—
Coronene (XXIV)	19,410	18, 32	9.4
d. Hydrocarbons with Five-Membered Rings			
Fluoranthene (XXVI)	18,510	18	—
3,4-Benzfluoranthene (XXVII)	19,190[e]	—	2.0[e]
2,13-Benzfluoranthene (XXVIII)	18,970[e]	—	—
Fluorene (XXIX)	23,910	38	—
1,2-Benzfluorene (XXX)	20,080	18	—
2,3-Benzfluorene (XXXI)	20,080	18	—
3,4-Benzfluorene (XXXII)	19,340[e]	—	1.1[e]
Truxene (XXXIII)	22,470	21	—
Isotruxene (XXXIV)	19,720	39	—
e. Condensed Polyphenyls			
Triphenylene (XXXV)	23,250	18	15.9
1,2:3,4:5,6:7,8-Tetrabenzanthracene (XXXVI)	20,550	18	8.9[e]
1,2:3,4:6,7:12,13-Tetrabenzpentacene (XXXVII)	19,570[e,f]	—	8.4[e]
5,6:8,9:14,15:17,18-Tetrabenzheptacene (XXXVIII)	19,160[e,f]	—	3.3[e,f]
1,2:6,7-Dibenzpyrene (XXXIX)[g]:			
1,12:2,3:10,11-Tribenzperylene (XL)	18,620[e,f]	—	3.3[e,f]
1,2:3,4:5,6:10,11-Tetrabenzanthanthrene (XLI)	18,590[e,f]	—	4.4[e,f]
1,12:2,3:4,5:6,7:8,9:10,11-Hexabenzcoronene (XLIII)	17,540[e,f]	—	3.8[e,f]

[a] All data obtained from low-temperature measurements.

[b] The names and numbering are those used by E. Clar, "Polycyclic Hydrocarbons." Academic Press, New York, 1964.

[c] References are cited only when the complete spectrum of the substance in question is reproduced either as tables or as diagrams in the publications indicated.

[d] When no other reference is given in this column the values have been taken from Table 5, the footnotes to which give the appropriate literature citations.

[e] From M. Zander, unpublished work (1965).

[f] Measurements made in trichlorobenzene. [g] See Section c of the table.

In addition to the acenes and phenes just discussed, there are many more hydrocarbons of the most varied structural types, the phosphorescence spectra of which have been similarly investigated. The positions of the phosphorescence bands of many angular hydrocarbons such as chrysene, 3,4-benzphenanthrene, 1,2-benzchrysene, picene, etc., in EPA at 77°K, have been reported by Clar and Zander,[18] and Table 9 contains

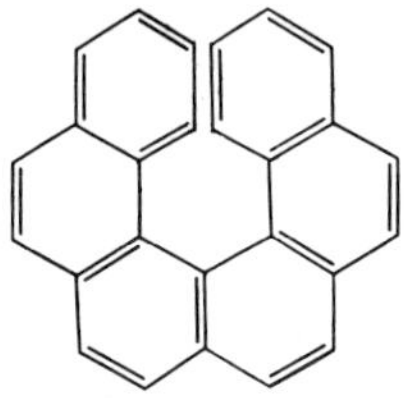

XV 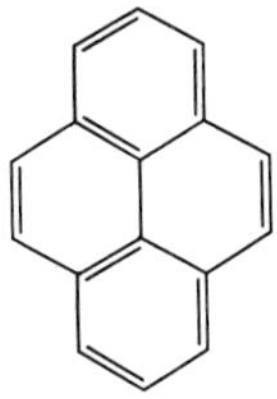 XVI 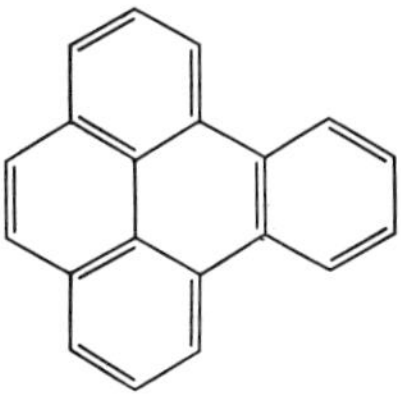 XVII

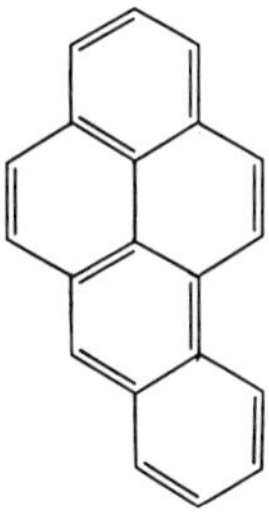

XVIII 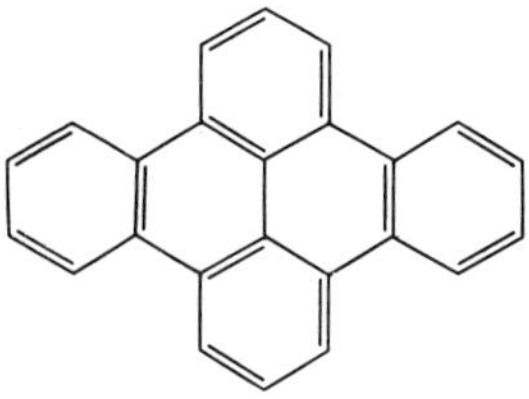 XIX 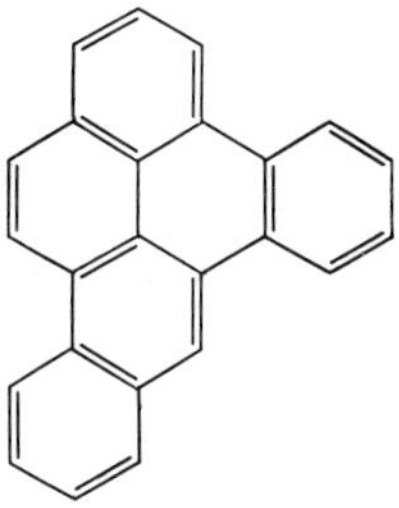 XX

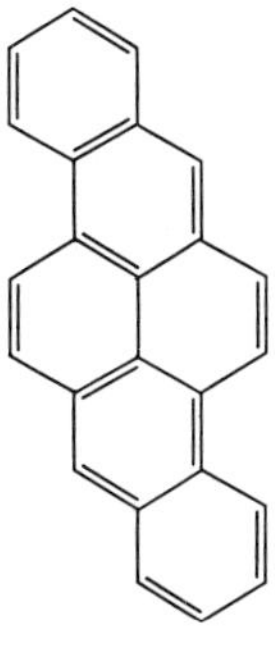

XXI 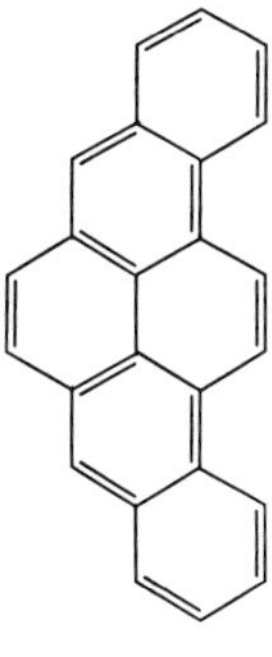 XXII

a selection of the measurements given by them. O'Dwyer *et al.*[26] have measured the phosphorescence spectrum of the interesting compound hexahelicene (XV).

In the pyrene series the phosphorescence spectra of pyrene (XVI),[14,18,27] 1,2-benzpyrene (XVII),[18] 3,4-benzpyrene (XVIII),[28] 1,2:6,7-dibenzpyrene (XIX),[18] 1,2:4,5-dibenzpyrene (XX),[29] 3,4:8,9-dibenzpyrene (XXI),[30] and 3,4:9,10-dibenzpyrene (XXII)[30] are known. Those of pyrene, and 1,2-benzpyrene have also been studied in a crystalline paraffin matrix by Shpol'skii *et al.*[31] For the latter, 150 lines were measured at 4°K.

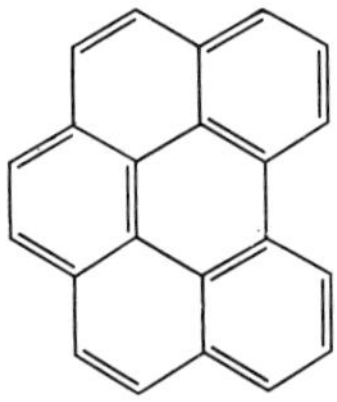

XXIII

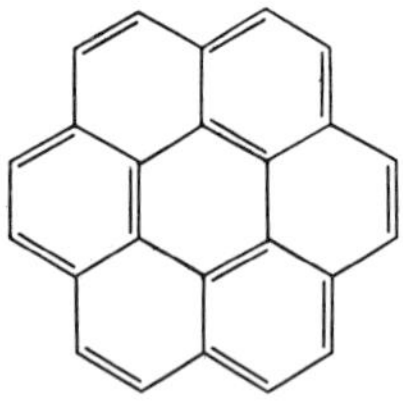

XXIV

The phosphorescence of perylene is not known. However, the hydrocarbons 1,12-benzperylene (XXIII)[18] and coronene (XXIV),[18,32] which are formally related to perylene (see Fig. 6 and Table 9), have been investigated. Shpol'skii and Klimova,[33] who have studied the phosphorescence spectrum of coronene in a matrix of hexane and heptane,

[26] M. F. O'Dwyer, M. A. El-Bayoumi, and S. J. Strickler, *J. Chem. Phys.* **36**, 1395 (1962).

[27] D. S. McClure, *J. Chem. Phys.* **17**, 905 (1949); A. A. Iljina and E. W. Shpol'skii, *Nachr. Akad. Wiss., Ud. SSR, Physik. Ser.* **15**, 585 (1951).

[28] B. Muel and M. Hubert-Habart, *J. Chim. Phys.* **55**, 377 (1958).

[29] A. A. Iljina and E. W. Shpol'skii, *Nachr. Akad. Wiss., Ud. SSR, Physik. Ser.* **15**, 585 (1951).

[30] B. Muel and M. Hubert-Habart, *Proc. 4th Intern. Meeting Mol. Spectry., Bologna*, p. 647 (1959).

[31] E. W. Shpol'skii and E. Girdzhiyauskaite, *Opt. i Spektroskopiya* **4**, 620 (1958); E. W. Shpol'skii, L. A. Klimova, and R. J. Personov, *ibid.* **13**, 341 (1962).

[32] D. S. McClure, *J. Chem. Phys.* **17**, 905 (1949); E. Bowen and B. Brocklehurst, *J. Chem. Soc.* p. 4320 (1955); J. Czekalla and K. J. Mager, *Z. Elektrochem.* **66**, 65 (1962).

[33] E. Shpol'skii and L. A. Klimova, *Izv. Akad. Nauk. SSSR, Ser. Fiz.* **23**, 23 (1959).

conclude from the vibrational structure, which is distinctly different from that of the fluorescence spectrum, that coronene is substantially deformed in the triplet state. De Groot and van der Waals[34] as well as Nieman and Tinti[35] have come to the same conclusion about other compounds.

Among aromatic hydrocarbons with one or more five-membered rings the following have been examined (see Table 9): acenaphthene (XXV),[36] fluoranthene (XXVI),[18] 3,4-benzofluoranthene (XXVII),[37] 2,13-benzofluoranthene (XXVIII),[37] fluorene (XXIX),[38] and the three benzofluorenes (XXX),[18] (XXXI),[18] and (XXXII),[37] as well as truxene (XXXIII)[21] and isotruxene (XXXIV).[39]

Biphenyl,[3,18,40] p-terphenyl,[39] 1,3,5-triphenylbenzene,[18] and 2,2′-dinaphthyl[18] are examples of hydrocarbons of the diaryl type whose phosphorescence spectra are known.

A specially interesting class of polycyclic aromatic hydrocarbons whose phosphorescence characteristics have been investigated is that opened up by Clar *et al.*,[41] the condensed polyphenyls. Though triphenylene (XXXV), their simplest representative, has been known for a long time, most of the condensed polyphenyls XXXVI to XLIII have only recently been synthesized. Clar and his colleagues have postulated[41] that in these compounds the π electrons are not uniformly distributed over the whole system, but are effectively localized in groups of six (Robinson's aromatic sextets) in the benzene rings that are marked with

34 M. de Groot and J. van der Waals, *Mol. Phys.* **6**, 545 (1963).

35 G. Nieman and D. Tinti, communicated to the American Institute of Physics Meeting, 1965.

36 A. Zmerli, *J. Chem. Phys.* **34**, 2130 (1961); W. W. Trussow and P. O. Tepljakow, *Fiz. Zh.* **8**, 1353 (1963); *Opt. i Spektroskopiya* **16**, 52 (1964); *ibid.* p. 27.

37 M. Zander, unpublished data (1965).

38 D. S. McClure, *J. Chem. Phys.* **17**, 905 (1949); R. C. Heckman, *J. Mol. Spectry.* **2**, 27 (1958); Y. Kanda, R. Shimada, K. Hanada, and S. Kajigaeshi, *Spectrochim. Acta* **17**, 1268 (1961).

39 K. F. Lang, M. Zander, and E. A. Theiling, *Chem. Ber.* **93**, 321 (1960).

40 Y. Kanda, R. Shimada, and Y. Sakai, *Spectrochim. Acta* **17**, 1 (1961).

41 E. Clar and M. Zander, *J. Chem. Soc.* p. 1861 (1958); E. Clar and C. T. Ironside, *Proc. Chem. Soc.* p. 150 (1958); E. Clar, C. T. Ironside, and M. Zander, *J. Chem. Soc.* p. 142 (1959); E. Clar, G. S. Fell, and M. H. Richmond, *Tetrahedron* **9**, 96 (1960); E. Clar and A. McCallum, *ibid.* **20**, 507 (1964); M. Zander, *Chem. Ber.* **92**, 2744 (1959); comprehensive account: E. Clar, "Polycyclic Hydrocarbons," Vol. I, Academic Press, New York, p. 37ff. 1964.

circles. This model of the condensed polyphenyls has since been corroborated by quantum mechanical investigations.[42]

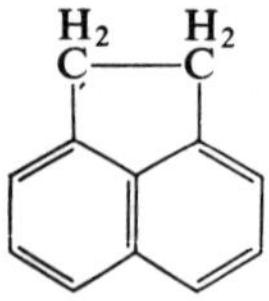

H_2 H_2 C—C

XXV 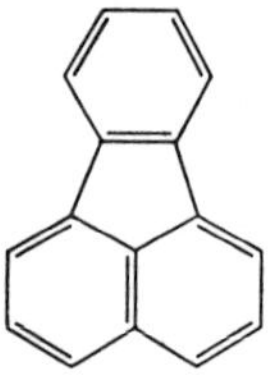 XXVI 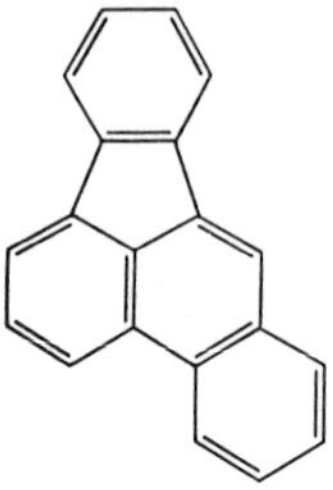 XXVII 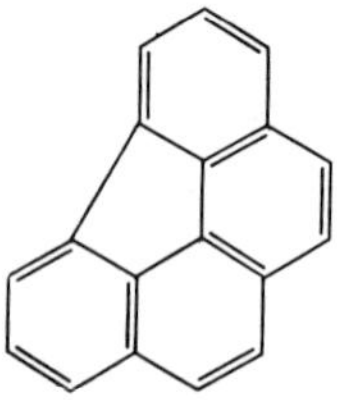 XXVIII

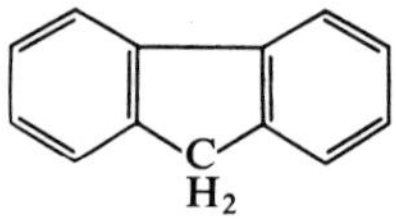

C H_2 C H_2 C H_2

XXIX 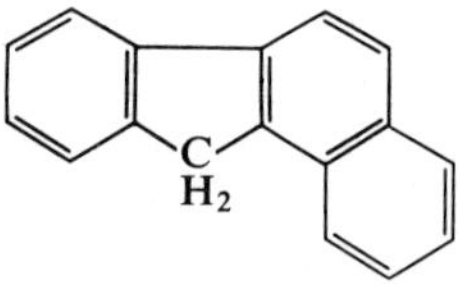 XXX 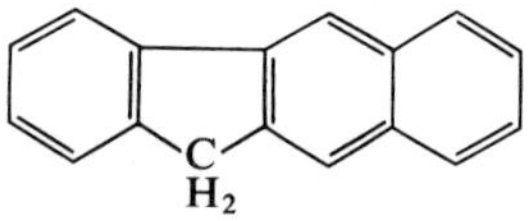 XXXI

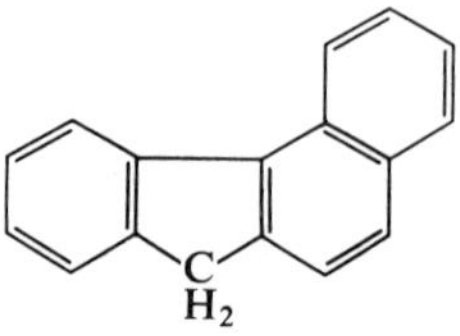

C H_2

XXXII

CH_2 H_2C C H_2

CH_2 H_2 C H_2C

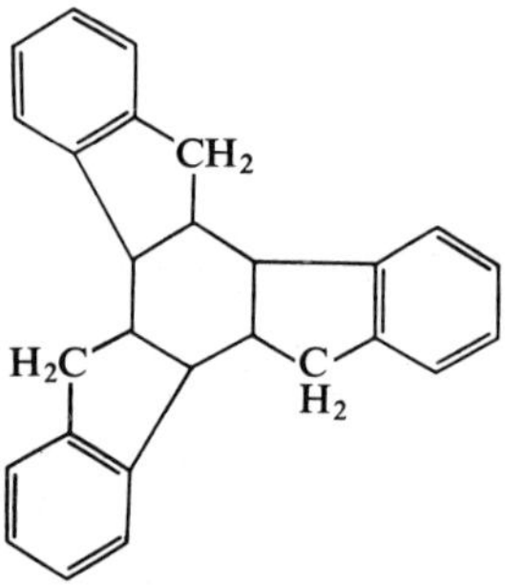

XXXIII 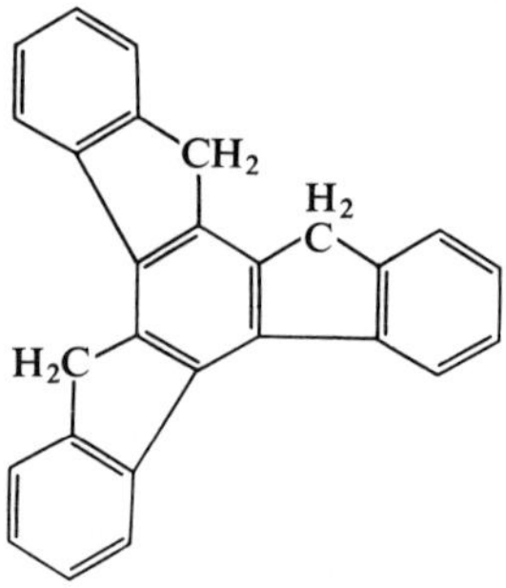 XXXIV

[42] R. Paunczand A. Cohen, *J. Chem. Soc.* p. 3288 (1960).

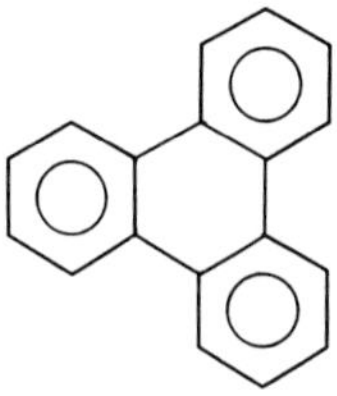

XXXV

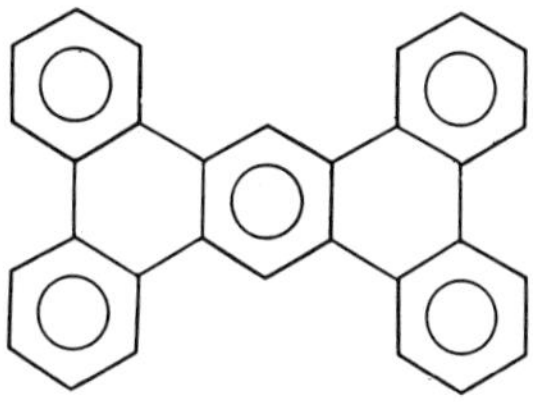

XXXVI

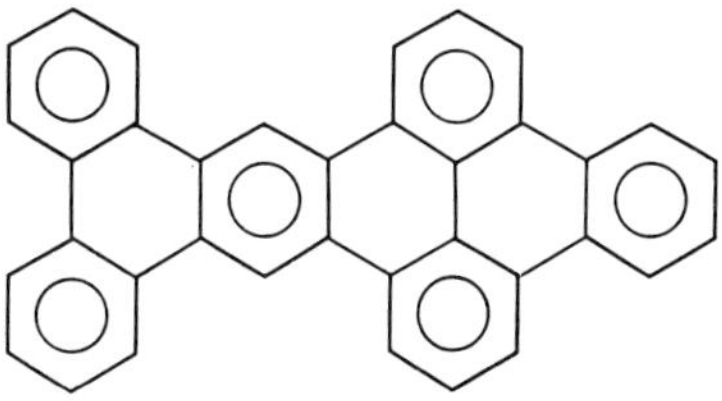

XXXVII

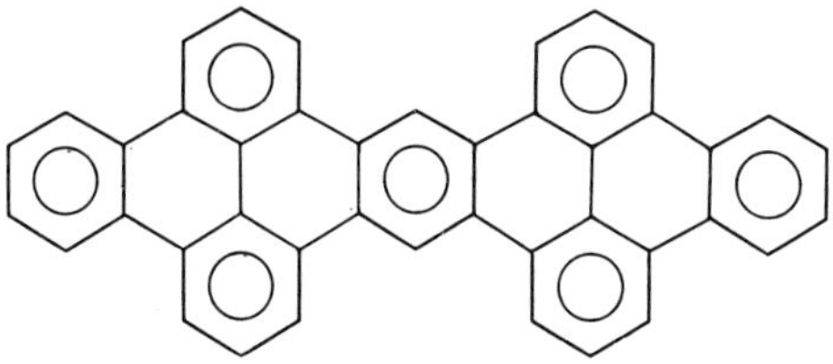

XXXVIII

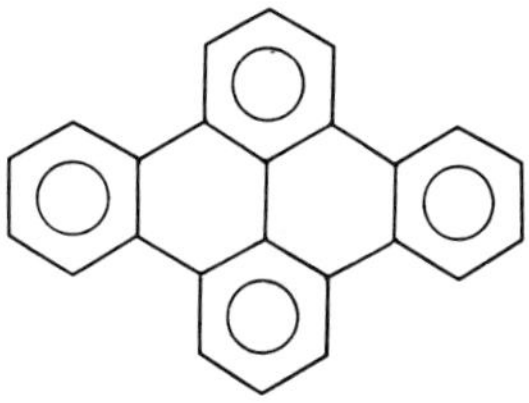

XXXIX

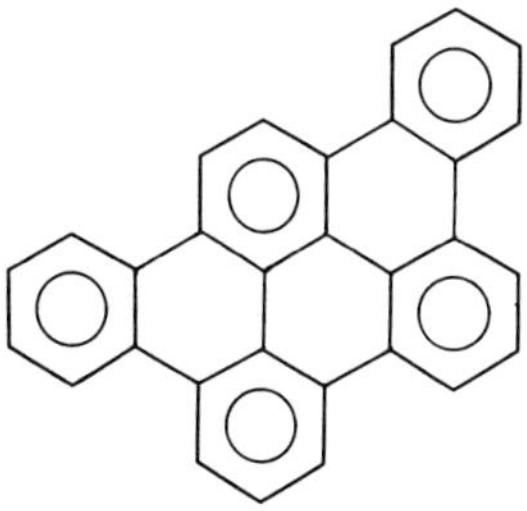

XL

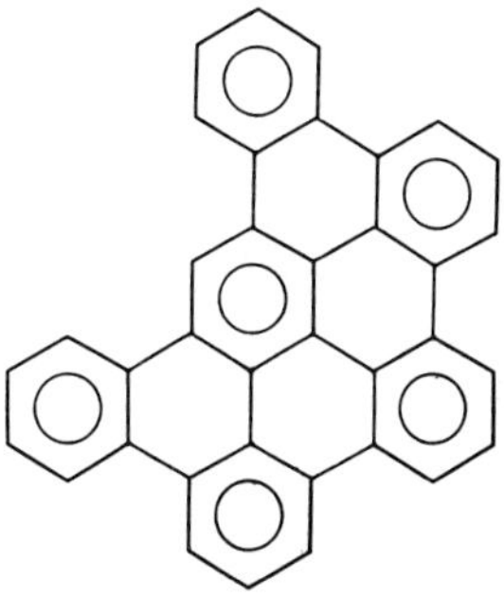

XLI

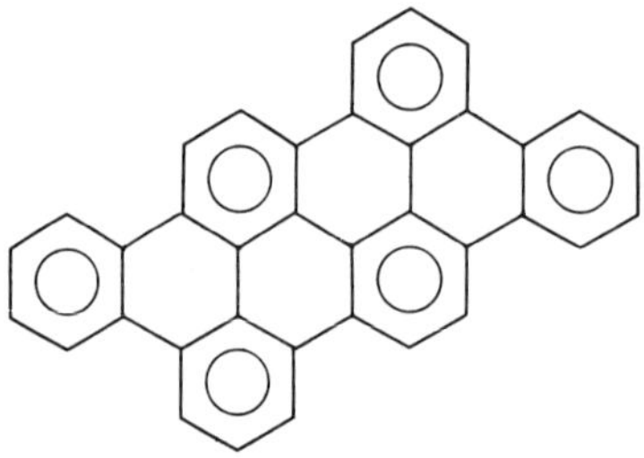

XLII

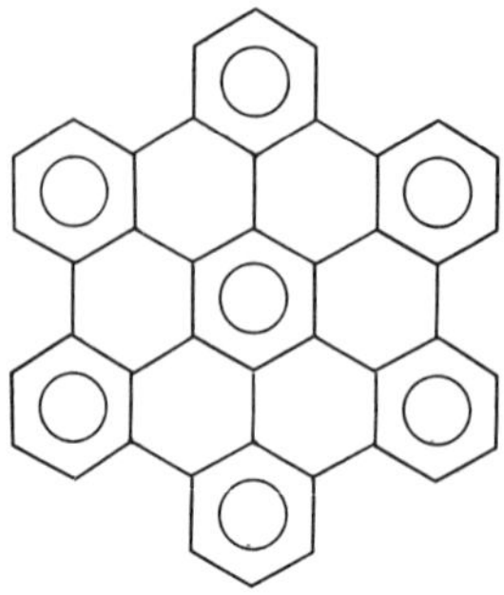

XLIII

Spectroscopically the condensed polyphenyls are distinguished in two important respects. First, their UV spectra lie at considerably shorter wavelengths than those of isomeric hydrocarbons of other structures; thus the band of longest wavelength of the compound XXXVII, which is formally derived from the blue substance pentacene, lies at 399 mμ and that of the highly condensed hexabenzocoronene XLIII is found at 444 mμ. Second, they give very intense phosphorescence with remarkably long lifetimes.

The mean phosphorescent lifetimes of the condensed polyphenyls XXXV, XXXVI, XXXVII, and XXXIX have been measured in EPA and in 1,2,4-trichlorobenzene.[37] They lie between 8 seconds for XXXIX and 16 seconds for XXXV (triphenylene) in EPA and around 3–4 seconds in trichlorobenzene as a consequence of the "external heavy atom effect" of the solvent (see Section 1.2). Because of the sparing solubility of hydrocarbons XXXVIII, XL, XLI, and XLIII, their phosphorescent lifetimes could be determined only in trichlorobenzene. They are between 3 and 4 seconds for all the compounds investigated and it must be assumed that they are at least twice as great in EPA. If the unusually long lifetime of triphenylene is attributed to some additional effective symmetry prohibition (see Section 1.3) operative only on this substance, then an average value for the lifetime of the condensed polyphenyls is about 10 seconds in EPA. Among the aromatic hydrocarbons so long a phosphorescent lifetime is found only for benzene and coronene.

As the phosphorescence transition is shifted increasingly toward the red, i.e., as the term difference $S_0 - T_1$ becomes smaller, the lifetime also usually gets rapidly smaller because of the increasing importance of the radiationless $T_1 \dashrightarrow S_0$ transition. In contrast the phosphorescent lifetime of the condensed polyphenyls is remarkably independent of the position of this transition.

All known condensed polyphenyls phosphoresce very intensely, but systematic measurements of the quantum yields have not yet been made. There are, however, a few quantitative results available to support the qualitative conclusions. Thus the ratio ϕ_p/ϕ_f of the quantum yield in phosphorescence ϕ_p to that in fluorescence ϕ_f amounts, according to Parker and Hatchard,[43] to about 0.8 for phenanthrene, but for triphenylene, the phosphorescence spectrum of which lies in much the same region, it is about 5 and the researches of McClure *et al.*[44] even give about 12. According to our own measurements, though they are only rough approximations, ϕ_p/ϕ_f is about 0.5 for 1,2-benzpyrene, but about 3 for the condensed polyphenyl XXXIX (1,2:6,7-dibenzpyrene). Hence one can suppose on the basis of the qualitative and the small amount of quantitative evidence just submitted that the particular structure of the condensed polyphenyls gives rise to specially high probabilities of intersystem crossing. Table 9 includes a collection of phosphorescence data for compounds of this type.

There exists a series of general relationships between the phosphorescence behavior, the UV absorption, and the structures of aromatic hydrocarbons. It can usually be expected that the quantum yield ϕ_p of the phosphorescence will be all the greater the smaller the extinction coefficient of the longest-wavelength singlet–singlet absorption transition and the shorter the wavelength of the phosphorescence. The dependence on the moment of the first S–S transition follows from the fact that the smaller this is, the greater is the ratio of the probabilities of intersystem crossing and fluorescence and therefore of ϕ_p/ϕ_f. The dependence on the wavelength of the phosphorescence transition arises from the fact that as the term difference $S_0 - T_1$ becomes smaller so the radiationless deactivation of the triplet state increases.[45] From these relationships intense phosphorescence may be expected especially for those hydrocarbons whose para absorption band (^1L_a) lies at low wavelength

[43] C. A. Parker and C. G. Hatchard, *Analyst* **87**, 664 (1962).
[44] E. H. Gilmore, G. E. Gibson, and D. S. McClure, *J. Chem. Phys.* **20**, 829 (1952); for correction, see *ibid.* **23**, 399 (1955).
[45] G. W. Robinson and R. P. Frosch, *J. Chem. Phys.* **38**, 1187 (1963).

(because of the relationship between phosphorescence and para absorption) and whose longest-wavelength *S–S* transition corresponds to an α band (^1L_b) ($\epsilon = 10^2$–10^3). Conversely, hydrocarbons will only phosphoresce weakly if the first band of their UV spectrum is a long-wavelength para band. Anthracene, with its para band at 375 mμ and log $\epsilon = 3.87$, is an example.

The mean phosphorescent lifetimes of aromatic hydrocarbons lie between a few milliseconds and several seconds. Compared with other types of compounds such as carbonyl or halogenated compounds most aromatic hydrocarbons display an unusually long phosphorescence and this implies that spin–orbit coupling takes place only to a slight extent in these molecules. This is also revealed by ESR measurements on naphthalene.[46] Various quantum mechanical investigations on the subject of the spin–orbit coupling in aromatic hydrocarbons have been published and have led, in some cases, to quite contradictory conclusions.[47] The possibility has been discussed that singlet–triplet mixing takes place not only through spin–orbit coupling but also by other mechanisms.

Benzene has been found specially interesting, partly because it is the simplest member of the aromatic series, and partly because of its very long mean and natural triplet lifetimes. McClure *et al.* estimated 7 seconds for the mean phosphorescent lifetime and 21 seconds as the upper limit for the natural lifetime of the lowest triplet state.[44] Robinson *et al.*[48] concluded from low temperature measurements (at 4.2°K) on perdeuterobenzene that the natural lifetime of the triplet state of benzene is 26 seconds. A value more than 10 times as great as this has been assumed by Craig *et al.*[49] on the basis of *S–T* absorption measurements, but the conclusions relying on this value have not remained unchallenged.[48] To summarize, one may say that the question of the magnitude of the natural lifetime of the lowest triplet state of benzene is still not satisfactorily cleared up.

Mean phosphorescent lifetimes of many aromatic hydrocarbons have been measured. Some values are given in Table 5 and further figures may

[46] C. A. Hutchison and B. W. Mangum, *J. Chem. Phys.* **34**, 908 (1961).
[47] D. S. McClure, *J. Chem. Phys.* **17**, 905 (1949); **20**, 686 (1952); M. Mizushima and S. Koide, *ibid.* p. 765; H. F. Hameka and J. L. Oosterhoff, *Mol. Phys.* **1**, 358 (1958).
[48] M. R. Wright, R. P. Frosch, and G. W. Robinson, *J. Chem. Phys.* **33**, 934 (1960).
[49] D. P. Craig, J. M. Hollas, and G. W. I. King, *J. Chem. Phys.* **29**, 974 (1958).

be found in Table 9. Reference has already been made several times to the fall of phosphorescent lifetime with increasing red shift of the phosphorescence transition (see Section 1.4). Thus the mean lifetime of the acenes alters by a factor of about 1000 between benzene and tetracene as the position of the transition changes (0,0 phosphorescence band of benzene = 29,500 cm^{-1}, $\tau_0 = 7$ seconds; for tetracene, 0,0 band = 10,250 cm^{-1}, $\tau_0 = 0.005$ second; data for tetracene from *S–T* absorption).[15]

The phosphorescence spectra of numerous homologs of aromatic hydrocarbons have been measured. That of toluene[3, 50] has been the

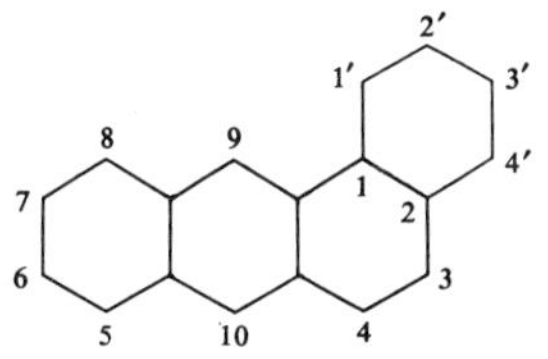

XLIV

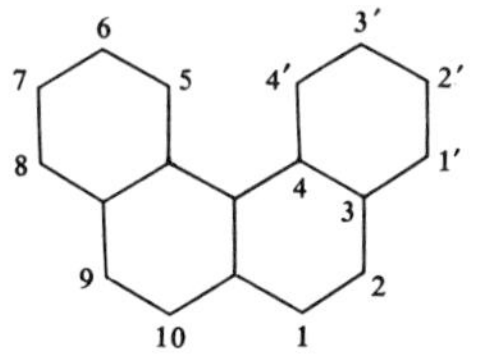

XLV

subject of several investigations, including one by Kanda and Shimada,[51] who have studied it in a cyclohexane matrix and have given a vibrational analysis for it.

Examinations of the phosphorescence of *o*-, *m*-, and *p*-xylene,[51, 52] mesitylene,[51, 53] durene,[53, 54] hexamethylbenzene,[53] and of both mono-methylnaphthalenes[3] are available.

Hirshberg[55] and Moodie and Reid[56] have investigated the phosphorescence characteristics of the 12 isomeric monomethyl-1,2-benzanthracenes (XLIV) and of the six isomers of monomethyl-3,4-benzphenanthrene (XLV) in connection with studies of the carcinogenic properties of aromatic hydrocarbons. As would be supposed, the spectra of the homologs resemble those of the parent compounds. Nevertheless

[50] P. P. Dikun and B. Ya. Sveshnikov, *Zh. Eksperim. i. Teor. Fiz.* **19**, 1000 (1949); Y. Kanda and H. Sponer, *J. Chem. Phys.* **28**, 798 (1958).
[51] Y. Kanda and R. Shimada, *Spectrochim. Acta* **17**, 279 (1961).
[52] P. Pesteil and A. Zmerli, *Ann. Phys.* **10**, 1079 (1955).
[53] H. Sponer and Y. Kanda, *J. Chem. Phys.* **40**, 778 (1964).
[54] Y. Meyer and R. Astier, *J. Phys. Radium* **24**, 1089 (1963); J. Czekalla and K. J. Mager, *Z. Elektrochem.* **66**, 65 (1962).
[55] Y. Hirshberg, *Anal. Chem.* **28**, 1954 (1956).
[56] M. M. Moodie and C. Reid, *J. Chem. Phys.* **22**, 252 (1954).

between the homologs and the parent compounds and between the individual homologs there occur small differences in the positions and relative intensities of the bands and these can be useful in identifying these compounds analytically. Among the homologs of benzanthracene, the 9- and 10-monomethyl and also the 9,10-dimethyl compounds have definitely shorter lifetimes than any other isomers.[55] It is therefore interesting that these are the most strongly carcinogenic of the homologs of benzanthracene.

The 0,0 band of the phosphorescence of 3,4-benzphenanthrene has been reported by Hirshberg[55] as occurring at 496 mμ (in alcohol–methanol–ether, 8:2:1), and all the monomethyl benzphenanthrenes except the 5 compound have 0,0 bands between 501 and 504 mμ. That of 5-methyl-3,4-benzphenanthrene (XLV), however, occurs at 512 mμ. Here the phosphorescence clearly reveals a steric effect—overcrowding of the hydrogen atoms that are found at the 4′ position and in the methyl group.

Zander[37] has measured the phosphorescence spectra of monomethyl- and monoethylcoronene.[57] In position, the bands correspond very closely with those of coronene, but in the distribution of intensities there is a definite difference from that of the parent substance in that the 0,0 bands are relatively intense. This is another pointer to the conclusion that the extreme weakness of the 0,0 phosphorescence band of coronene must be attributed to a symmetry prohibition additional to the genuine intercombination prohibition.

2.2. Substitution Products of Aromatic Hydrocarbons

Although the phosphorescent lifetimes and quantum yields of aromatic hydrocarbons are greatly altered by halogen substitution, the effect on the positions of the transitions is relatively small. The halogen monosubstitution products of naphthalene have been investigated particularly thoroughly,[1] and in Table 10 the positions of their 0,0 bands in glassy solid solutions at 77°K have been collected for all these fluoro, chloro, bromo, and iodo compounds. As will be seen, halogen substitution shifts the phosphorescence toward longer wavelengths (with the one exception of 2-fluoronaphthalene), and substitution in the 1 position has a greater

[57] E. Clar, B. A. McAndrew, and M. Zander, *Tetrahedron*, **23**, 985 (1967).

[1] J. Ferguson, T. Iredale, and J. A. Taylor, *J. Chem. Soc.* p. 3160 (1954).

effect than in the 2 position. Although the relative intensity distribution of the naphthalene spectrum remains almost unchanged by fluorine substitution, it is considerably affected by the heavier halogen atoms. Complete phosphorescence spectra of the monohalogenonaphthalenes (except the iodine derivatives) have been given by Ferguson, Iredale, and Taylor.[1]

TABLE 10 Phosphorescence of Monohalogenonaphthalenes[a, b]

	1-Isomer		2-Isomer	
Halogen	0,0 Band (cm^{-1})	Δ (cm^{-1})	0,0 Band (cm^{-1})	Δ (cm^{-1})
F	21,062	184	21,268	−22
Cl	20,645	601	21,069	177
Br	20,652	594	21,036	210
I	19,000(?)	2246(?)	21,040	206

[a] Δ = displacement of 0,0 band from 0,0 band of naphthalene (21,246 cm^{-1}).

[b] Data for 1-Iodonaphthalene are from R. V. Nauman, Thesis, University of California (1947), and for 2-Iodonaphthalene, G. N. Lewis and M. Kasha, *J. Am. Chem. Soc.* **66**, 2100 (1944); all the rest are from J. Ferguson, T. Iredale, and J. A. Taylor, *J. Chem. Soc.* p. 3160 (1954). All measurements were made at 77°K in light petroleum (except for the iodonaphthalenes, which were in EPA).

The introduction of several halogen atoms into an aromatic hydrocarbon causes the red shift of the phosphorescence to increase, its magnitude depending, among other things, on the positions of the substituents. The influence of multiple substitution of halogens on the wavelengths of the transitions of naphthalene and anthracene is shown by Table 11. If the α positions of anthracene are successively occupied by chlorine atoms, the red shift averages about 200 cm^{-1} per chlorine. As would be expected the effect of substitution at the 9 and 10 positions of anthracene is somewhat greater.[2]

The ratio of the quantum yields of fluorescence to phosphorescence is definitely altered by halogen substitution in naphthalene but the total quantum yield remains approximately the same. This means that with

[2] M. R. Padhye, S. P. McGlynn, and M. Kasha, *J. Chem. Phys.* **24**, 588 (1956).

naphthalene the introduction of the heavy atoms has only a small influence on the frequencies of the radiationless transitions from the lowest singlet and triplet states to the ground state. Benzene behaves quite differently. Not only the quantum yield of the fluorescence but also that of the phosphorescence is appreciably smaller for the halogenobenzenes than for unsubstituted benzene,[3] and this points to a great

TABLE 11 Phosphorescence of Di-, Tri-, and Tetrahalogenated Naphthalenes and Anthracenes[a, b]

Compound	0,0 Band (cm^{-1})	Δ (cm^{-1})
1,5-Dibromonaphthalene	20,086	1160
2,6-Dibromonaphthalene	20,806	440
1-Chloroanthracene	14,732	195
1,5-Dichloroanthracene	14,568	359
1,4,5,8-Tetrachloroanthracene	14,155	772
1,10-Dichloroanthracene	14,128	799
1,5,10-Trichloroanthracene	13,800	1127
9,10-Dichloroanthracene	14,150	700
9,10-Dibromoanthracene	14,060	790

[a] Δ = displacement from naphthalene (21,246 cm^{-1}) or anthracene (14,927 cm^{-1}).

[b] Results for the dibromonaphthalenes from J. Ferguson, T. Iredale, and J. A. Taylor, *J. Chem. Soc.* p. 3160 (1954); the remainder from M. R. Padhye, S. P. McGlynn, and M. Kasha, *J. Chem. Phys.* **24**, 588 (1956). The values for 9,10-dichloro- and 9,10-dibromoanthracene are derived from *S–T* absorption measurements at room temperature; all the rest from phosphorescence measurements at 77°K, the halogenoanthracenes in EPA and the halogenonaphthalenes in light petroleum.

increase in the radiationless deactivation of the excited states. Ferguson and Iredale[4] have reported on the extremely weak phosphorescence of iodobenzene as well as of *o*-, *m*-, and *p*-chloroiodobenzene.

The introduction of an —OH, —SH, or —NH_2 group into an aromatic hydrocarbon also causes a red shift of the phosphorescence spectrum. Table 12 contains the positions of the 0,0 bands and the lifetimes of a number of phenols. For the phosphorescence of the dihydroxy phenols

[3] E. H. Gilmore, G. E. Gibson, and D. S. McClure, *J. Chem. Phys.* **20**, 829 (1952).

[4] J. Ferguson and T. Iredale, *J. Chem. Soc.* p. 2959 (1953).

hydroquinone, resorcinol, and pyrocatechol, see Zudin.[5] Information concerning the positions of the phosphorescence transitions of a number of aromatic amines such as aniline, diphenylamine, 1- and 2-naphthylamine, 2-aminofluorene, and others has been given by Lewis and Kasha.[6]

The phosphorescence of aromatic nitro compounds is of the n–π^* type.[7] Usually these substances show no fluorescence, whereas their

TABLE 12 Phosphorescence of Phenols[a]

Compound	0,0 Band (cm^{-1})	Lifetime (sec.)
Phenol	28,500	2.9
1-Naphthol	20,500	1.9
2-Naphthol	21,100	1.3
Thio-2-naphthol	20,800	0.28
4-Phenanthrol	21,150	2.1

[a] Results for 4-phenanthrol are unpublished measurements by M. Zander (1965). For all other compounds band positions are according to G. N. Lewis and M. Kasha, *J. Am. Chem. Soc.* **66**, 2100 (1944), and lifetimes according to D. S. McClure, *J. Chem. Phys.* **17**, 905 (1949). All measurements were made at 77°K in EPA.

phosphorescences attain high quantum yields (0.8–1.0); their phosphorescent lifetimes are short (e.g., 1-nitronaphthalene, 0.049 second; naphthalene, 2.6 seconds), and in the vibrational structure of their spectra there appears the Raman frequency characteristic of the aromatic nitro group (ca. 1450 cm^{-1}).

Lewis and Kasha[6] have published the phosphorescence transitions of numerous aromatic nitro compounds, and Corkill and Graham-Bryce[8] have reported on those of the 10 isomeric dinitronaphthalenes and

[5] A. A. Zudin, *Izv. Akad. Nauk SSSR, Ser. Fiz.* **23**, 142 (1959); *Chem. Abstr.* **53**, 11,989 (1959).
[6] G. N. Lewis and M. Kasha, *J. Am. Chem. Soc.* **66**, 2100 (1944).
[7] M. Kasha, *Radiation Res.* Suppl. 2, 243 (1960).
[8] J. M. Corkill and I. J. Graham-Bryce, *J. Chem. Soc.* p. 3893 (1961).

McClure has given some results for the lifetimes of aromatic nitro compounds.[9] Table 13 contains a selection of their data.

Some interesting results have been found for the nitroanilines and the nitronaphthylamines, which have been investigated by Foster *et al.*[10] and by Corkill and Graham-Bryce,[8] respectively. These compounds show *either* fluorescence *or* phosphorescence. Perhaps Foster's measurements ought to be extended, since his apparatus was only able to record

TABLE 13 Phosphorescence of Aromatic Nitro Compounds[a]

Compound	0,0 Band (cm^{-1})	Lifetime (sec.)
Nitrobenzene	21,100	—
1-Nitronaphthalene	18,800	0.05
1,2-Dinitronaphthalene	17,300	—
1,3-Dinitronaphthalene	19,230	—
1,4-Dinitronaphthalene	17,800	—
1,5-Dinitronaphthalene	19,125	0.11
2-Nitronaphthalene	19,550	—
2,6-Dinitronaphthalene	18,850	—
2-Nitrofluorene	20,600	0.13
4-Nitrobiphenyl	20,500	0.08

[a] Data for the nitronaphthalenes are from J. M. Corkill and I. J. Graham-Bryce, *J. Chem. Soc.* p. 3893 (1961); for all other compounds, from G. N. Lewis and M. Kasha, *J. Am. Chem. Soc.* **66**, 2100 (1944). Phosphorescent lifetimes are from D. S. McClure, *J. Chem. Phys.* **17**, 905 (1949). All measurements were made at 77°K in EPA or ethanol.

phosphorescences that had lifetimes greater than 0.1 second. Of the three isomeric nitroanilines, only the 1,4 isomer phosphoresces but the other two fluoresce. Again, of the 14 isomeric nitronaphthylamines three phosphoresce (1-nitro-2-naphthylamine and 4- and 5-nitro-1-naphthylamine), all the others showing fluorescence exclusively. From these results it seems likely that there is a connection between the luminescence properties of the compounds and the conjugation of the two

[9] D. S. McClure, *J. Chem. Phys.* **17**, 905 (1949).

[10] R. Foster, D. L. Hammick, G. M. Hood, and A. C. E. Sanders, *J. Chem. Soc.* p. 4865 (1956).

substituents. In 12 of the 17 nitroarylamines investigated conjugation between the NO_2— and NH_2— groups is prevented by steric interference of substituents in ortho positions or is impossible on electronic grounds since meta quinoid systems would be involved. Eleven of these 12 compounds fluoresce; only one phosphoresces. Among the remaining five nitroarylamines, in which conjugation is complete between the substituents, three phosphoresce and two fluoresce. If a further substituent is introduced into 4-nitroaniline in an ortho position to the

TABLE 14 Phosphorescence of Nitroarylamines[a]

Compound	0,0 Band (cm^{-1})
4-Nitroaniline	19,400
2-Methyl-4-nitroaniline	19,150
2,6-Dimethyl-4-nitroaniline	18,950
4-Nitro-1-naphthylamine	17,400
5-Nitro-1-naphthylamine	17,100
1-Nitro-2-naphthylamine	17,350

[a] Data for the nitroanilines are from R. Foster, D. L. Hammick, G. M. Hood, and A. C. E. Sanders, *J. Chem. Soc.* p. 4865 (1956); for the nitronaphthylamines, from J. M. Corkill and I. J. Graham-Bryce, *J. Chem. Soc.* p. 3893 (1961). All measurements were made in ethanol at low temperatures.

NO_2— group the phosphorescence disappears, but a substituent ortho to the NH_2— group has no effect on the luminescent behavior. From this Foster *et al.*[10] have derived a model of 4-nitroaniline in the triplet state. Table 14 gives the positions of the phosphorescence transitions of several of the compounds investigated.

In the wider sense there belong also to this chapter the aromatic carbonyl compounds: aldehydes, ketones, carboxylic acids, and so on. The majority of the aromatic aldehydes and ketones that have been investigated show, like the nitro compounds, n–π^* phosphorescence.[7] The lifetimes of these phosphorescences are, as a rule, extremely short, and so the emissions are frequently observable not only in rigid solutions at low temperatures, but also in liquid solutions at room temperature, in the vapor phase, and in the crystalline state, as has already been referred to elsewhere (see Section 1.1). Sandros and Backstrom[11] have observed the phosphorescence of benzil, anisil, and biacetyl in oxygen-free solutions in benzene at room temperature; the phosphorescence decay period of ca. 5×10^{-5} second reaches approximately that of the long-lived fluorescence, as has been found for naphthalene, for example.[12] The singlet–triplet absorption spectra of aromatic carbonyl compounds

TABLE 15 Phosphorescence of Aromatic Aldehydes, Ketones, and Carboxylic Acids[a]

Compound	0,0 Band (cm^{-1})	Lifetime (sec.)
Benzaldehyde	25,200[b]	1.6×10^{-3} [c]
1-Naphthaldehyde	19,900[b, d]	ca. 1[d]
Acetophenone	26,000[b]	7.5×10^{-3} [c]
Benzophenone	24,400[b]	5.0×10^{-3} [c]
Anthrone	25,000[d]	
1-Benzoic acid	27,200[b]	2.5[e]
2-Naphthoic acid	20,900[b]	2.5[e]

[a] All measurements were made in EPA at 77°K.
[b] G. N. Lewis and M. Kasha, *J. Am. Chem. Soc.* **66**, 2100 (1944).
[c] A. Martinez, *Compt. Rend.* **255**, 491 (1962).
[d] R. Shimada and L. Goodman, *J. Chem. Phys.* **43**, 2027 (1965). According to these authors 1-naphthaldehyde displays π–π^* phosphorescence.
[e] D. S. McClure, *J. Chem. Phys.* **17**, 905 (1949).

with n–π^* phosphorescence, like benzaldehyde, acetophenone, and benzophenone, also show interesting deviations from the usual behavior. As Kanda *et al.*[13] have found, the intensity of the *S–T* absorption of these compounds is not raised by the presence of oxygen. Reference has already been made to the high quantum yields of aromatic carbonyl compounds with n–π^* phosphorescence and to the fact that they do not usually fluoresce (see Section 1.3).

[11] H. L. J. Backstrom and K. Sandros, *Acta Chem. Scand.* **14**, 48 (1960).
[12] M. Kasha and R. V. Nauman, *J. Chem. Phys.* **17**, 516 (1949).
[13] Y. Kanda, H. Kaseda, and T. Matumura, *Spectrochim. Acta* **20**, 1387 (1964).

Phosphorescence spectra have been measured for numerous aromatic carboxylic acids,[14] such as benzoic, salicylic, phthalic, and gallic acids, and for acid derivatives such[15] as *p*-aminobenzoic acid, *o*- and *p*-bromobenzoic acid, etc. Similar measurements have also been reported for the nitriles[16] (benzonitrile, *o*- and *p*-dicyanobenzene, *o*-, *m*-, and *p*-toluonitrile).

Table 15 gives phosphorescence data for some aromatic aldehydes, ketones, and carboxylic acids.

2.3. Quinones

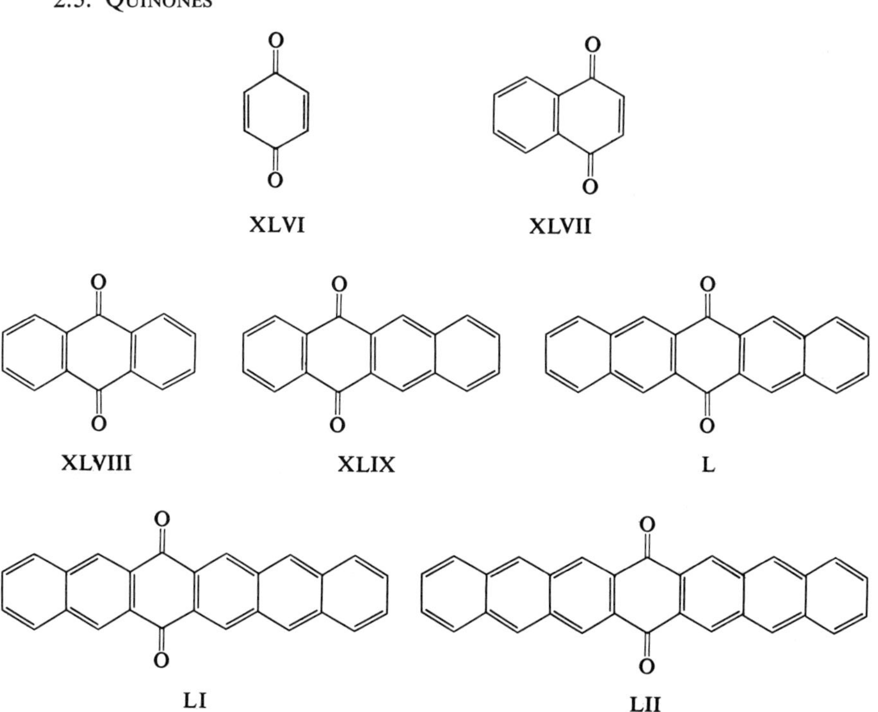

[14] B. A. Pyatnitskii, *Dokl. Akad. Nauk SSSR* **109**, 503 (1956); *Chem. Abstr.* **51**, 9330a (1957); *Izv. Akad. Nauk SSSR, Ser. Fiz.* **23**, 135 (1959); *Chem. Abstr.* **53**, 11,991b (1959).

[15] P. A. Teplyakov, *Izv. Vysshikh Uchebn. Zavedenii, Fiz.* p. 135 (1959); *Chem. Abstr.* **53**, 19,567g (1959).

[16] K. Takei and Y. Kanda, *Spectrochim. Acta* **18**, 1201 (1962).

For the quinones, of which *p*-benzoquinone (XLVI), naphtho-1,4-quinone (XLVII), and the monoquinones XLVIII–LII have been relatively well investigated, there exist, as for the aromatic aldehydes and ketones (see Section 2.2), two types of excited states, π,π^* states and n,π^* states.† As is well known, the latter are derived from the electronic configuration of the ground state by the promotion of an electron from a nonbinding atomic orbital (such as a lone-pair electron of the carbonyl oxygen) into a π^* orbital. The luminescence behavior of the quinones XLVI–LII is effectively determined, if it is known whether the lowest singlet excited state is of π,π^* or n,π^* type.

The bands of longest wavelength that are observed in the UV spectra of hexacene-6,15-quinone (LI) and heptacene-7,16-quinone (LII) are intense π–π^* bands; they are found at 454 mμ (log $\epsilon = 4.08$) for quinone LI and 460 mμ (log $\epsilon = 4.38$) for quinone LII in trichlorobenzene. That the band shift is so small in passing from the hexacene quinone to the heptacene is in agreement with Hartmann's "part system rule."[1]

Zander[2] has found that in solution at room temperature LI and LII show an intense fluorescence, the spectrum of which is the mirror image of that of the longest-wavelength π–π^* absorption band group. If the fluorescence is investigated in mixed crystals with phenanthrene, naphthalene, or trichlorobenzene at −196°C the spectrum, which is only vaguely structured in liquid solution at room temperature, is split up into several narrow bands. The fluorescence spectra of the quinones LI and LII show the same dependence on solvent as the longest-wavelength π,π^* absorption band, i.e., they are shifted toward longer wavelengths with increasing dielectric constant of the solvent. There can, from these results, remain no doubt that the lowest singlet excited states of these hexacene and heptacene quinones are π,π^* states. This is also, at least for the heptacene quinone, in agreement with quantum mechanical calculations of Pullman and Diner.[3]

Phosphorescence could definitely not be demonstrated for hexacene and heptacene quinones. Of course if the emissions were very weak and occurred at long wavelengths their observation might be very difficult.

† It should be noted that n–π^* and π–π^* denote transitions; but n,π^* and π,π^*, states.

[1] H. Hartmann and E. Lorenz, *Z. Naturforsch.* **7a**, 360 (1952).

[2] M. Zander, *Naturwissenschaften* **53**, 404 (1966).

[3] B. Pullman and S. Diner, *J. Chim. Phys.* **55**, 212 (1958).

This situation is considered in the term scheme reproduced in Fig. 14a, with which the observed luminescence properties of the quinones LI and LII are comparable. However, other possible explanations are conceivable.

The bands of longest wavelength in the UV spectrum of the quinones XLII–L are weak ($\epsilon < 10^2$) n,π^* bands of the CO group and lie between ca. 420 and 470 mμ.[1] Like the vast majority of compounds, the lowest singlet excited states of which are of n,π^* type, the quinones show no

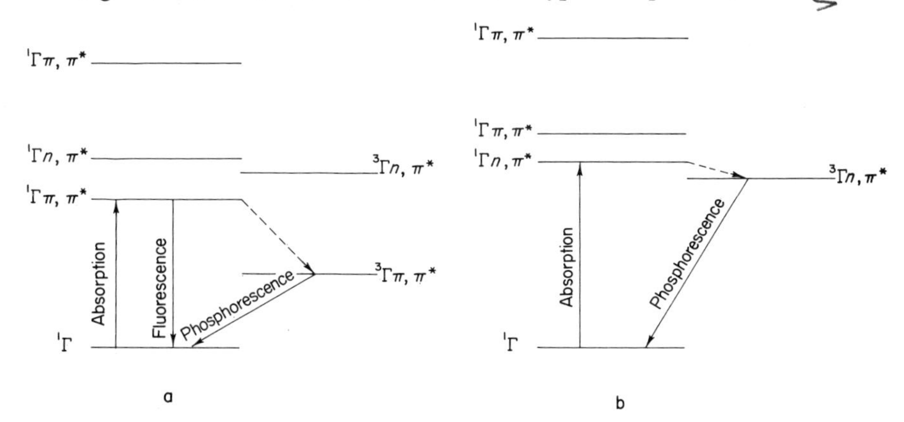

Fig. 14. Term scheme for quinones. (a) Heptacene quinone type; (b) Anthraquinone type.

measurable fluorescence either at room temperature or in rigid solution at low temperature. On the other hand, for these quinones well-structured intense phosphorescence spectra are observed that correspond to the transition from the n,π^* triplet state to the ground state. They occur between about 450 and 520 mμ and depend only slightly on the sizes of the molecules.[2] The phosphorescence spectrum of tetracene-5,12-quinone (XLIX) in EPA is reproduced in Fig. 15 as an example. Further splitting of the bands can be observed in iodobenzene solution. The luminescent behavior of the quinones XLVII–L corresponds with the term scheme shown in Fig. 14b. The small singlet–triplet splitting of the n,π^* configuration is worth noting; its order of magnitude is 1000–3000 cm^{-1}.

The question of the position and properties of the lowest π,π^* triplet states of these quinones is of interest. In intensity and vibrational structure their longest-wavelength π,π^* absorption bands are comparable with the para bands (the L_a bands) of the aromatic hydrocarbons. We ought, therefore, to assume that the singlet–triplet separations of the first π,π^* states of the quinones correspond with those observed for the para bands, that is, have the approximate magnitude of 10,000 cm^{-1}.

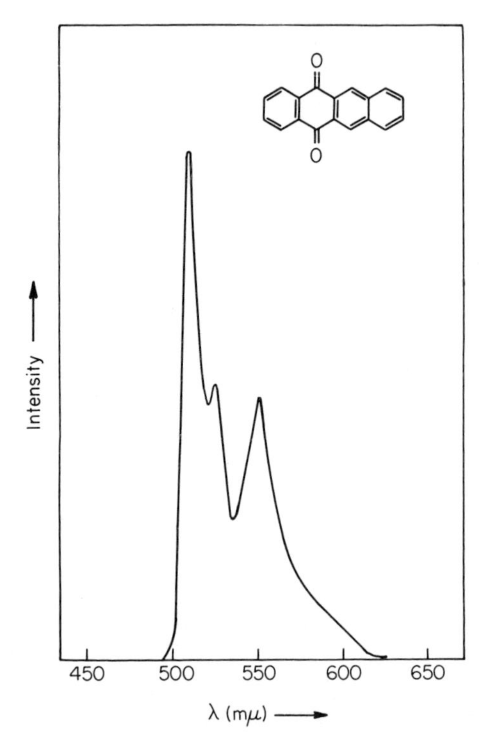

Fig. 15. Phosphorescence spectrum of tetrazene-5,12-quinone (XLIX) in EPA at 77° K. [According to M. Zander, *Z. Elektrochem.* **71**, 424 (1967).]

From the position of the first π–π^* absorption transition (ca. 25,000 cm^{-1}) for the tetracene and pentacene quinones there results then a π,π^* triplet state at about 15,000 cm^{-1}. This requires, however, that the π,π^* triplet state for these compounds lie considerably lower than the observed n,π^* triplet state (ca. 19,500 cm^{-1}). If this argument is accepted it would seem that the quinones provide an exception to

Kasha's rule,[4] according to which in any molecule only the prevailing lowest state of a given multiplicity is capable of emission (see Section 1.1). From the experimental material so far available it is not possible to decide whether this is the case or whether the assumption made—equality of the singlet–triplet splitting of the π,π^* states in quinones and hydrocarbons—is incorrect.

In the matter of its structure, *p*-benzoquinone (XLVI) occupies a rather special position in relation to the quinones considered so far. In its UV spectrum n,π^* absorption is observed[1, 5] between about 500 and 280 mμ. This absorption has been investigated in the crystalline state [6] at 20°K as well as in the gas state.[7] Kanda *et al.*[8] accept a value of 18,682 cm^{-1} for the position of the n,π^* triplet state on the basis of a detailed analysis of the spectrum of the vapor.

2.4. Aromatic *N*-Heterocyclics

Pyridine, the simplest aromatic *N*-heterocyclic, shows no phosphorescence. Its *S–T* absorption spectrum has been measured by Evans.[1]

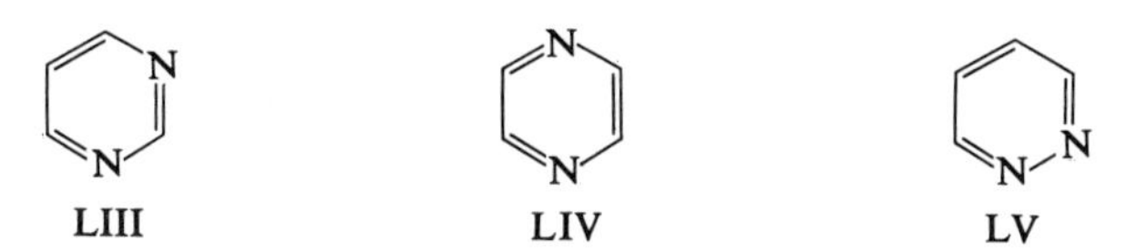

Of the monocyclic diazines, pyrimidine[2] (LIII) and pyrazine[3] (LIV) phosphoresce, whereas pyridazine[4] (LV) does not. The emission of the diazines is n–π^* and the quantum yields are high. The compounds show no measurable fluorescence either in the vapor state or in liquid or solid

[4] M. Kasha, *Discussions Faraday Soc.* **9**, 14 (1950).
[5] F. W. Kingstedt, *Compt. Rend.* **176**, 1550 (1923).
[6] J. Sidman, *J. Am. Chem. Soc.* **78**, 2363 (1956).
[7] T. Anno and A. Sado, *J. Chem. Phys.* **32**, 1602 (1960).
[8] Y. Kanda, H. Kaseda, and T. Matsumura, *Spectrochim. Acta* **20**, 1387 (1964).
[1] D. F. Evans, *J. Chem. Soc.* p. 3885 (1957).
[2] M. Krishna and L. Goodman, *J. Am. Chem. Soc.* **83**, 2042 (1961); R. Shimada, *Spectrochim. Acta* **17**, 30 (1961).
[3] L. Goodman and M. Kasha, *J. Mol. Spectry.* **2**, 58 (1958); R. Shimada, *Spectrochim. Acta* **17**, 14 (1961).
[4] M. A. El-Sayed, *J. Chem. Phys.* **36**, 573 (1962).

solution. This behavior points to a very high rate for the intersystem crossing process. El-Sayed[5] has shown that intersystem crossing between states of different orbital character ($^1n,\pi^* \leftrightarrow {}^3\pi,\pi^*$ or $^1\pi,\pi^* \leftrightarrow {}^3n,\pi^*$) takes place with considerably greater rate than between

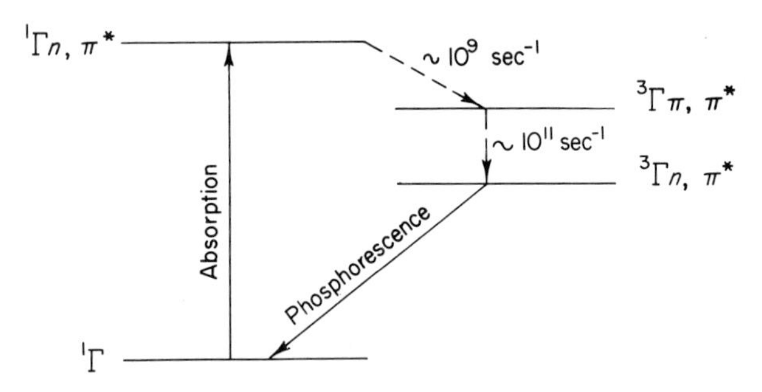

Fig. 16. Term scheme for aromatic *N*-heterocyclics (diazines).

states with the same orbital character ($^1n,\pi^* \leftrightarrow {}^3n,\pi^*$ or $^1\pi,\pi^* \leftrightarrow {}^3\pi,\pi^*$). From this it follows according to El-Sayed that the term scheme shown in Fig. 16 can be accepted for the phosphorescent diazines. The intersystem crossing process $^1n,\pi^* \rightarrow {}^3\pi,\pi^*$ has a high rate (ca. 10^9 per second) and so, too, has the subsequent internal conversion from this π,π^* triplet state to the n,π^* triplet state by which the radiation is emitted. Hence the n,π^* triplet state becomes densely populated.

Paris and colleagues[6] have reported on the phosphorescence of *sym*-triazine (LVI), but in *sym*-tetrazine (LVII)[7] and its dimethyl derivative[8]

[5] M. A. El-Sayed, *J. Chem. Phys.* **38**, 2834 (1963); see also the comprehensive review by S. K. Lower and M. A. El-Sayed, *Chem. Rev.* **66**, 199 (1966).
[6] J. P. Paris, R. C. Hirt, and R. G. Schmitt, *J. Chem. Phys.* **34**, 1851 (1961).
[7] M. Chowdhury and L. Goodman, *J. Chem. Phys.* **36**, 548 (1962).
[8] M. Chowdhury and L. Goodman, *J. Chem. Phys.* **38**, 2979 (1963).

the rare case of an n,π^* fluorescence has been observed; another example has also been found by Lippert[9] in 9,10-diazaphenanthrene. An explanation of these n,π^* fluorescences of aromatic *N*-heterocyclics has been given by El-Sayed.[5]

The luminescent behavior of the bicyclic quinoline (LVIII) depends in an interesting way on the solvent.[10] In nonpolar solvents only phosphorescence appears, with no fluorescence. By analogy with many other examples it must, therefore, be concluded that the lowest singlet excited state in a nonpolar solvent is an n,π^* state. In polar solvents fluorescence is observed as well as phosphorescence. Clearly in a polar solvent the lowest singlet excited state is a π,π^* state. It has been known for a long

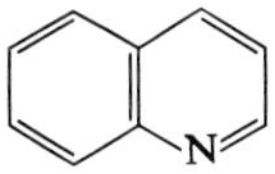

N N

LVIII 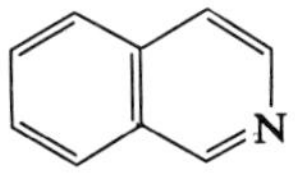 LIX

time that n,π^* absorption bands are displaced toward shorter wavelengths with increasing polarity of the solvent, and that π,π^* absorption bands move toward longer wavelengths. Clearly the lowest n,π^* and π,π^* singlet excited states of the quinoline molecule lie so close together that the sequence of the term series is reversed by change of solvent.

In contrast with the monocyclic azaaromatics, the phosphorescence of the bi- and polycyclic azaaromatics so far investigated is of π,π^* type. The phosphorescent lifetimes have the order of magnitude of seconds and in parts the spectra very closely resemble those of the corresponding aromatic hydrocarbons; for example, that of quinoline is very similar to that of naphthalene.

The phosphorescence spectrum of isoquinoline (LIX) is nearly the same as that of quinoline. It has been measured by Dörr and coworkers,[11] who have investigated these properties for a large number of aromatic *N*-heterocyclics.

[9] E. Lippert, *in* "Luminescence of Organic and Inorganic Materials" (H. P. Kallmann and G. M. Spruch, eds.), p. 274. Wiley, New York, 1962; E. Lippert and W. Voss, *Z. Physik. Chem.* (*Frankfurt*) [N.S.] **31**, 321 (1962).

[10] M. Kasha, *Radiation Res.* Suppl. 2, 243 (1960).

[11] F. Dörr and R. Müller, *Z. Elektrochem.* **63**, 1150 (1959); F. Dörr and H. Gropper, *ibid.* **67**, 193 (1963).

The phosphorescence spectra of several benzologs of quinoline—5,6-benzoquinoline (LX), 7,8-benzoquinoline (LXI), and phenanthridine (LXII)—have been measured.[12] They correspond very closely with that of phenanthrene. Vibrational analyses for LX and LXI have been given by Kanda and Shimada.[13]

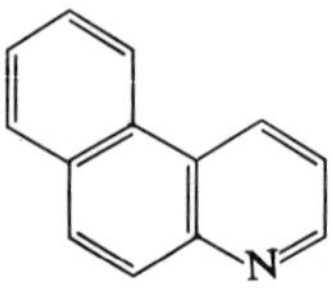

LX 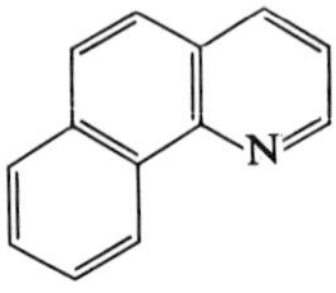 LXI 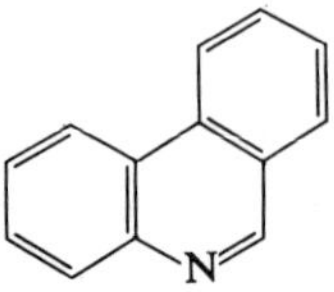 LXII

A few aromatic *N*-heterocyclics with two nitrogen atoms have also been studied, e.g., the phenanthrolines LXIII, LXIV, and LXV and benzoquinoxaline (LXVI),[12] which all correspond to phenanthrene.

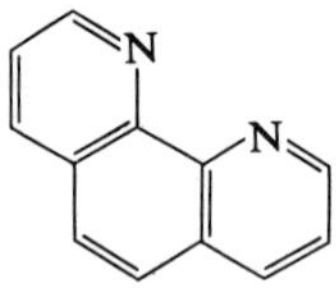

LXIII 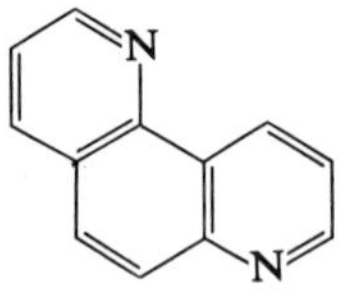 LXIV 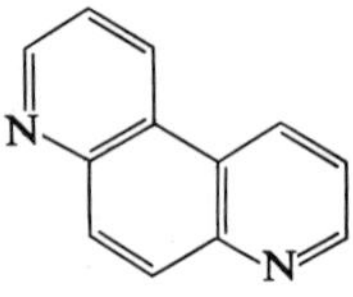 LXV

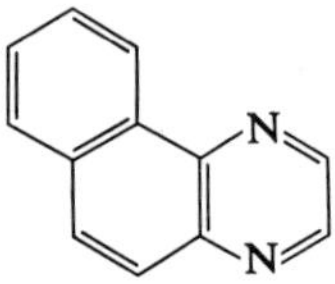

LXVI

A comparison of the phosphorescence spectra of dibenzoquinoxaline (LXVII) and of the corresponding hydrocarbon triphenylene (LXVIII) is interesting. Although the position of the transition in the two compounds is almost identical—the 0,0 bands lie at about 23,000 (LXVII) and 23,250 cm^{-1} (LXVIII)—the spectra are characteristically different in the distribution of the intensity. In the triphenylene spectrum the 0,0 band is extremely weak; in the dibenzoquinoxaline, on the other hand,

[12] H. Gropper and F. Dörr, *Z. Elektrochem.* **67**, 46 (1963).
[13] Y. Kanda and R. Shimada, *Spectrochim. Acta* p. 211 (1959).

this band is by far the most intense in the whole spectrum. From many examples it appears that, compared with the other bands, the 0,0 phosphorescence transition is scarcely altered if, in place of CH groups, one or more N atoms are built into the aromatic framework provided that the phosphorescence transition of the hydrocarbon is symmetry

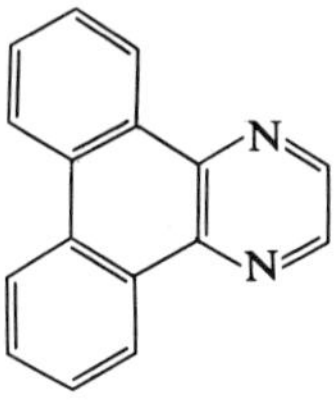

LXVII

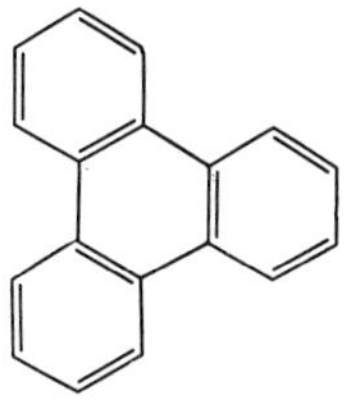

LXVIII

permitted. The entirely different properties of dibenzoquinoxaline and triphenylene support the supposition that the phosphorescence transition of the triphenylene is symmetry forbidden[11] (see Section 1.3).

The *T–S* transitions of several N-analogs of anthracene have been investigated. The *S–T* absorption spectra of acridine (LXIX), 1-azaanthracene (LXX), and 2-azaanthracene (LXXI) have been measured by Evans[14] together with many other aromatic *N*-heterocyclics. Phenazine (LXXII),[10] in contrast with anthracene, shows no fluorescence, and

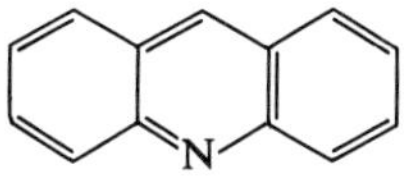

LXIX

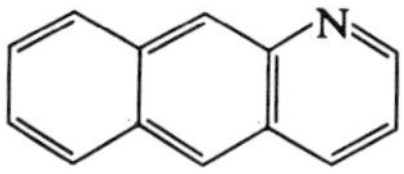

LXX

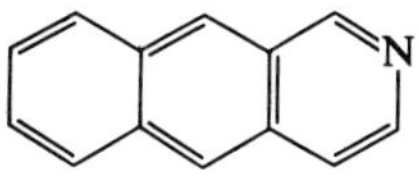

LXXI

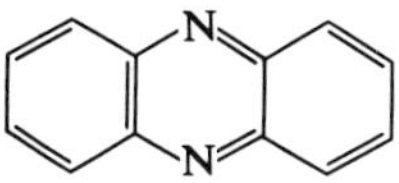

LXXII

though the phosphorescence of anthracene is extremely weak, that of phenazine has a quantum yield of ca. 1. All the observed *T–S* transitions of *N*-heterocyclic analogs of anthracene are of π,π^* type and agree in position very closely with the *T–S* transition of anthracene. The entirely

[14] D. F. Evans, *J. Chem. Soc.* p. 2753 (1959); *ibid.* p. 1351 (1957).

different ratio of the quantum yields for phosphorescence and fluorescence obtained on passing from anthracene to phenazine points to the participation of an n,π^* singlet excited state in the phenazine phosphorescence.

TABLE 16 Phosphorescence of Aromatic *N*-Heterocyclics

Compound	0,0 Band (cm^{-1})
Quinoline (LVIII)	21,700[a]
Isoquinoline (LIX)	21,200[b]
Naphthalene	21,300
5,6-Benzoquinoline (LX)	21,865[c]
7,8-Benzoquinoline (LXI)	21,790[c]
Phenanthridine (LXII)	22,050[d]
o-Phenanthroline (LXIII)	22,100[d]
m-Phenanthroline (LXIV)	22,150[d]
p-Phenanthroline (LXV)	22,200[d]
Benzoquinoxaline (LXVI)	21,000[d]
Phenanthrene	21,600
1-Azaanthracene (LXX)	15,070[e]
2-Azaanthracene (LXXI)	14,870[e]
Anthracene	14,850
1-Azapyrene (LXXIII)	16,860[f]
Pyrene	16,850

[a] G. N. Lewis and M. Kasha, *J. Am. Chem. Soc.* **66**, 2100 (1944).

[b] F. Dörr and H. Gropper, *Z. Elektrochem.* **67**, 193 (1963).

[c] Y. Kanda and R. Shimada, *Spectrochim. Acta* p. 211 (1959).

[d] H. Gropper and F. Dörr, *Z. Elektrochem.* **67**, 46 (1963).

[e] D. F. Evans, *J. Chem. Soc.* p. 2753 (1959). (*T–S* absorption).

[f] M. Zander and W. H. Franke, *Chem. Ber.* **99**, 1279 (1966).

Little is known concerning the phosphorescence spectra of *N*-heterocyclic analogs of peri-condensed hydrocarbons. That of 1-azapyrene (LXXIII)[15] agrees very closely with that of pyrene (LXXIV); their 0,0 bands lie at 16,860 and 16,850 cm^{-1}, respectively, and the *T–S* transition of 1-azapyrene has also been observed in absorption (at 16,930 cm^{-1}).[14]

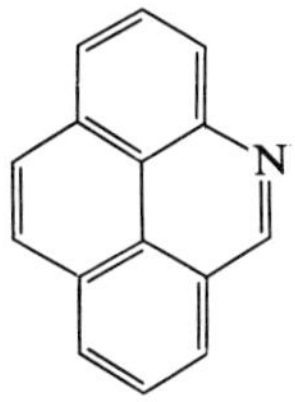

N

LXXIII

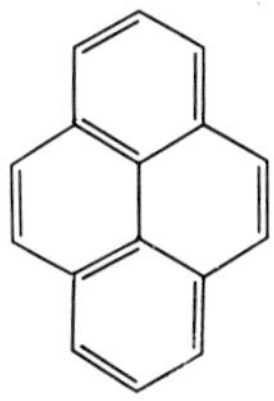

LXXIV

A selection of phosphorescence data for a few aromatic *N*-heterocyclics is given in Table 16.

2.5. Benzologs of Furan, Pyrrole, and Thiophene

With the simple five-membered-ring heterocyclics, furan, pyrrole, and thiophene, Fialkovskaya[1] and Heckman[2] were able to observe neither fluorescence nor phosphorescence. It may therefore be taken as certain that these compounds, if they emit at all, do so only extremely weakly. A "singlet–triplet absorption spectrum" that has been "observed"[3] for thiophene in the region of 315 mμ arises, as Evans[4] has shown, from impurities.

In contrast to the parent compounds the benzologs of furan, pyrrole, and thiophene produce intense phosphorescence of π,π^* type. Those of pyrrole have been investigated more thoroughly than the others.

The characteristic phosphorescence spectrum of indole (LXXV) is

[15] M. Zander and W. H. Franke, *Chem. Ber.* **99**, 1279 (1966).

[1] O. V. Fialkovskaya, *J. Phys. Chem.* (*USSR*) **11**, 533 (1938); *Acta Physicochim. USSR* **9**, 215 (1938).

[2] R. C. Heckman, *J. Mol. Spectry.* **2**, 27 (1958).

[3] M. R. Padhye and S. R. Desai, *Proc. Phys. Soc.* (*London*) **A65**, 298 (1952).

[4] D. F. Evans, *J. Chem. Soc.* p. 3885 (1957).

displayed in Fig. 17 and that of the next higher benzolog, carbazole (LXXVI), in Fig. 5. Zander[5] has similarly measured the mono- and dibenzocarbazoles (LXXVII–LXXXIV). It is revealed that the positions of the phosphorescence spectra of indole, carbazole, and the benzocarbazoles approximate closely those of naphthalene, phenanthrene, and the phenanthrene benzologs corresponding to LXXVII–LXXXIV.

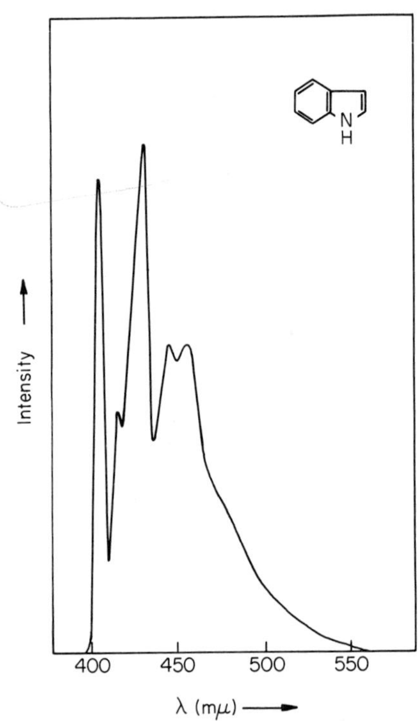

Fig. 17. Phosphorescence spectrum of indole (LXXV) in EPA at 77°K.

In Fig. 18 the positions of the 0,0 bands of the carbazoles and of the analogous aromatic hydrocarbons have been plotted. It will be seen that the positions of the transitions change in a very similar manner with the number of benzene nuclei and the direction of annellation in both series of compounds. With one exception the phosphorescence spectra of the

[5] M. Zander, *Chem. Ber.* **97**, 2695 (1964); *8th European Congr. Mol. Spectry., Copenhagen*, 1965 Report.

carbazoles lie at shorter wavelengths than those of the corresponding hydrocarbons. Their lifetimes are invariably longer. 3,4:5,6-Dibenzocarbazole (LXXXII), the one compound that does not conform to the relationship shown in Fig. 18, coincides remarkably well with the

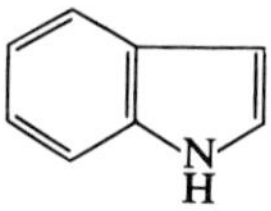

LXXV 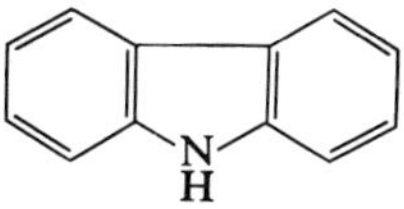 LXXVI

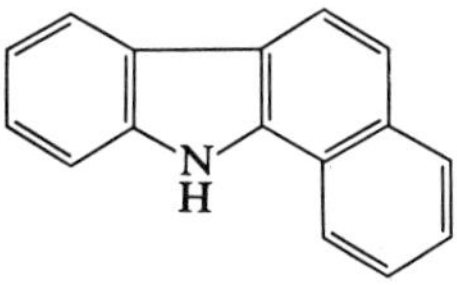

LXXVII 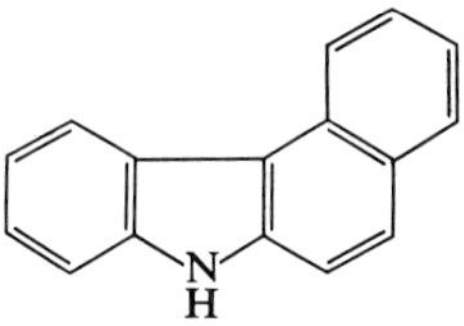 LXXVIII 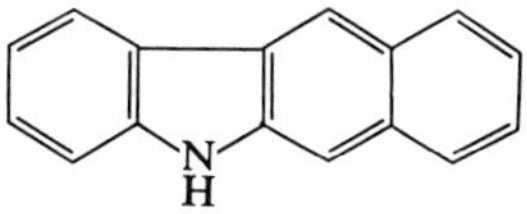 LXXIX

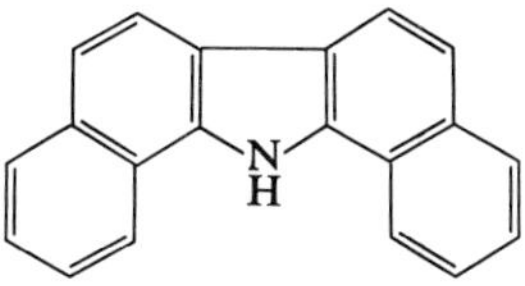

LXXX 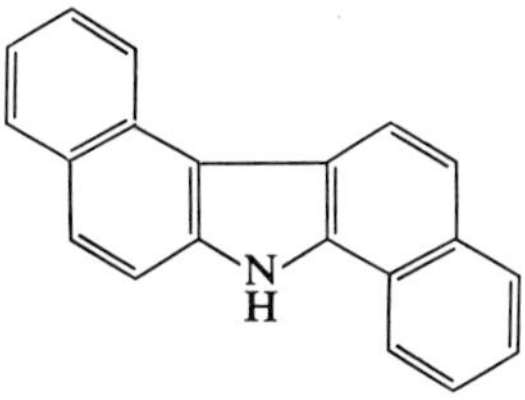 LXXXI 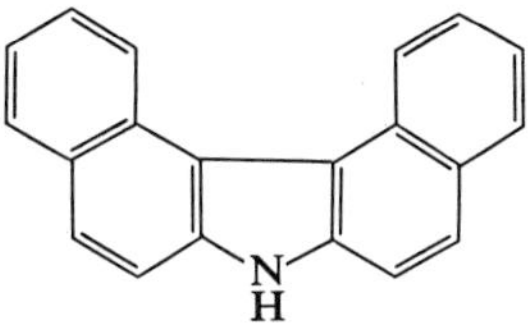 LXXXII

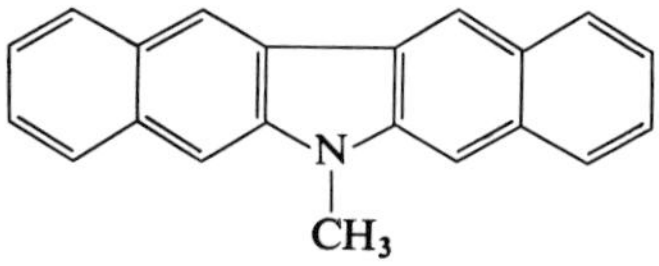

LXXXIII 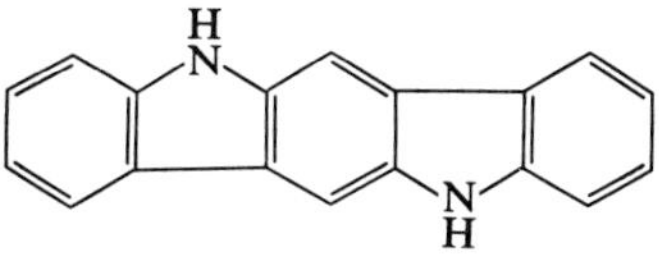 LXXXIV

corresponding hydrocarbon 3,4:5,6-dibenzophenanthrene not only in the position but also in the vibrational structure of its phosphorescence spectrum. The spectra of both compounds are reproduced in Fig. 19. Table 17 gives a collection of phosphorescence 0,0 bands of carbazoles and, for some of them, their phosphorescent lifetimes. It should be

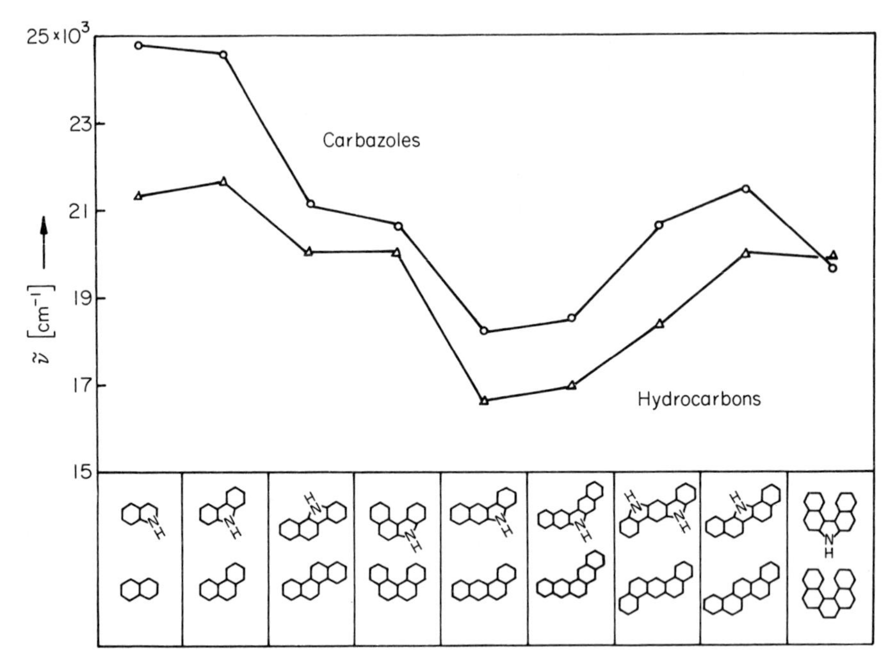

Fig. 18. Positions of the phosphorescence transitions of benzologs of carbazole and phenanthrene.

mentioned that the close analogy between carbazoles and benzologs of phenanthrene has also been shown to prevail in the UV spectra of numerous more highly annellated compounds.[6]

Phosphorescence data for 4,5-iminophenanthrene (LXXXV), which is, so far, the only known carbazole analog of a peri-condensed hydro-

NH

LXXXV LXXXVI

[6] M. Zander, *8th European Congr. Mol. Spectry., Copenhagen*, 1965 Report; M. Zander and W. H. Franke, *Chem. Ber.* **97**, 212 (1964); *ibid.* **98**, 588 (1965).

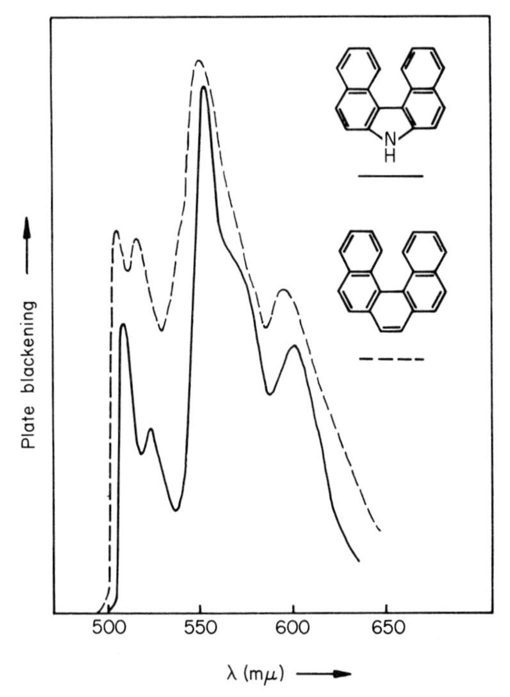

Fig. 19. Phosphorescence spectra of 3,4:5,6-dibenzocarbazole (LXXXII) (—) and of 3,4:5,6-dibenzophenanthrene (– –) in EPA at 77°K. [According to M. Zander, *Chem. Ber.* **97**, 2695 (1964).]

carbon, pyrene (LXXXVI), are also available.[7] The spectrum of 4,5-iminophenanthrene is displaced about 2400 cm^{-1} toward the violet compared with that of pyrene and its phosphorescent lifetime is considerably longer (1.5 second and 0.2 second). Its 0,0 band is given in Table 17.

Of the benzologs of thiophene, thionaphthene (LXXXVII) has been investigated thoroughly. Heckman[2] has measured its phosphorescence

S
LXXXVII

S
LXXXVIII

[7] M. Zander and W. H. Franke, *Chem. Ber.* **99**, 1279 (1966).

TABLE 17 Phosphorescence of Carbazoles[a]

Compound	0,0 Band (cm^{-1})	Lifetime (sec.)
Carbazole (LXXVI)[b]	24,510	—
9-Methylcarbazole[c]	24,450	7.8
9-Phenylcarbazole[c]	24,600	—
1,2-Benzocarbazole (LXXVII)[d]	21,050	3.6
3,4-Benzocarbazole (LXXVIII)[b]	20,700	3.3
2,3-Benzocarbazole (LXXIX)[b]	18,200	1.1
1,2:7,8-Dibenzocarbazole (LXXX)[b]	21,460	4.1
1,2:5,6-Dibenzocarbazole (LXXXI)[c]	20,660	3.0
3,4:5,6-Dibenzocarbazole (LXXXII)[b]	19,610	2.1
9-Methyl-2,3:6,7-Dibenzocarbazole (LXXXIII)[c]	18,380	—
Indolo-3′,2′:2,3-carbazole (LXXXIV)[b]	20,620	5.7
4,5-Iminophenanthrene (4-*H*-Benzo-(def)carbazole) (LXXXV)[e]	19,230	1.5

[a] All measurements were made at 77°K in EPA.
[b] From M. Zander, *Chem. Ber.* **97**, 2695 (1964).
[c] From M. Zander, unpublished work (1965).
[d] From M. Zander, *Erdoel Kohle* **19**, 278 (1966).
[e] From M. Zander and W. H. Franke, *Chem. Ber.* **99**, 1279 (1966).

TABLE 18 Phosphorescence of Benzologs of Thiophene and Furan[a]

Compound	0,0 Band (cm^{-1})	Lifetime (sec.)
Thionaphthene (LXXXVII)[b]	24,010	ca. 0.5
Diphenylene sulfide (LXXXIII)[b]	24,325	ca. 2.0
1,2-Benzodiphenylene sulfide (LXXXIX)[c]	20,830	0.9
2,3-Benzodiphenylene sulfide (XC)[c]	18,690	0.3
Diphenylene oxide (XCI)[b]	24,515	ca. 5.0
1,2-Benzodiphenylene oxide (XCII)[c]	21,280	4.0
2,3-Benzodiphenylene oxide (XCIII)[c]	19,230	1.4

[a] All measurements made in EPA at 77°K.
[b] From R. C. Heckman, *J. Mol. Spectry.* **2**, 27 (1958).
[c] From M. Zander, unpublished work (1965).

spectrum and given a vibrational analysis. The extremely low intensity of the band of shortest wavelength is noteworthy and Heckman has discussed it exhaustively. Using Kasha's method with ethyl iodide Padhye and Pabel[8] have studied the singlet–triplet absorption spectrum that is related as a mirror image to the phosphorescence spectrum of thionaphthene. The phosphorescence spectrum of diphenylene sulfide (LXXXVIII), which has also been examined by Heckman,[2] in contrast with that of thionaphthene, is shifted somewhat toward the red, but has a very similar vibrational structure, and the corresponding spectra of the benzodiphenylene sulfides (LXXXIX) and (XC) are likewise known.[9]

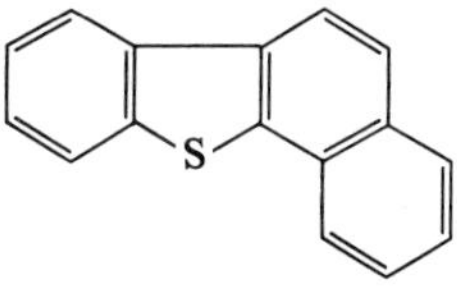

LXXXIX 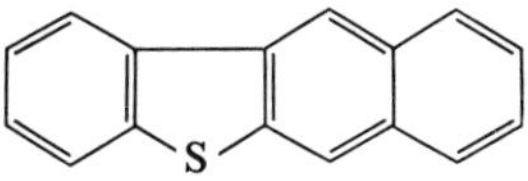 XC

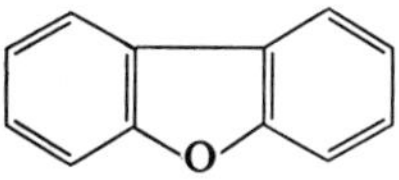

XCI

The phosphorescence spectrum of diphenylene oxide (XCI) has been measured at low temperatures, by Heckman[2] in EPA and by Kanda and colleagues[10] in a cyclohexane matrix. The spectrum reported by Kanda is extraordinarily rich in bands and a detailed vibrational analysis is given of it. Similar phosphorescence measurements are available for the benzodiphenylene oxides XCII and XCIII.[9]

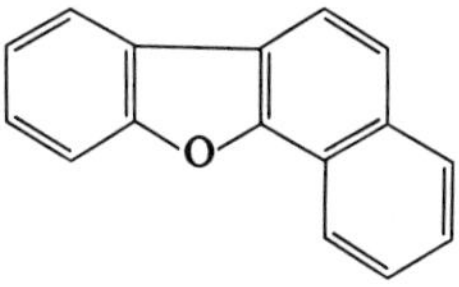

XCII 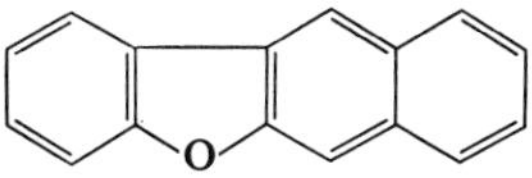 XCIII

A collection of phosphorescence data for benzologs of thiophene and furan is given in Table 18.

[8] M. R. Padhye and J. C. Pabel, *J. Sci. Ind. Res.* (*India*) **15B**, 206 (1956).
[9] M. Zander, unpublished data (1965).
[10] Y. Kanda, R. Shimada, H. Hanada, and S. Kajigaeshi, *Spectrochim. Acta* **17**, 1268 (1961).

2.6. Aliphatic Compounds

Most of the work that has appeared on triplet spectroscopy of organic compounds relates to aromatics. Very much less is known about aliphatic compounds.

The simplest organic compound that ought to show phosphorescence is ethylene. Actually, however, the phosphorescence reported for halogenated ethylenes by Lewis and Kasha[1] at about 4000 Å appears not to be genuine, and Potts,[2] who searched for the phosphorescence of tetramethylethylene, found none. Today it can be regarded as certain that ethylene itself and its simple derivatives show no measurable phosphorescence. The reasons for this are not really understood.

The position of the lowest triplet state of ethylene has, however, long been known from singlet–triplet absorption measurements. Snow and Allsopp[3] found in 1934 a weak absorption band at 2100 Å, which they immediately interpreted correctly as the singlet–triplet absorption. This work must have been the first in which a singlet–triplet absorption of an organic compound was observed and correctly identified. Measurements and calculations by later authors[2,4] confirmed the interpretation.

Evans[5] has measured the singlet–triplet absorption spectrum of butadiene by the oxygen method (see Section 1.2).

Lewis and Kasha[1] have investigated the phosphorescence properties of a long-chain polyene, lycopene ($C_{40}H_{56}$). An interesting feature is that the all-trans form showed scarcely any luminescence, though after conversion to a configuration in which cis groupings were present phosphorescence could be measured.

Beer[6] has reported on the phosphorescence of several polyacetylenes (in EPA at −170°C). Entirely aliphatic polyacetylenes as well as those with two phenyl substituents show phosphorescence that was found to occur at progressively greater wavelengths the greater the number of acetylene groups. Table 19 gives a selection of Beer's results.

[1] G. N. Lewis and M. Kasha, *J. Am. Chem. Soc.* **66**, 2100 (1944).

[2] W. J. Potts, Jr., *J. Chem. Phys.* **23**, 65 (1955).

[3] C. P. Snow and C. B. Allsopp, *Trans. Faraday Soc.* **30**, 93 (1934).

[4] W. E. Moffitt, *Proc. Phys. Soc.* (*London*) **63**, 1292 (1950); E. P. Carr and H. Stücklen, *J. Chem. Phys.* **4**, 760 (1936); L. W. Pickett, M. Muntz, and E. M. McPherson, *J. Am. Chem. Soc.* **73**, 4862 (1951).

[5] D. F. Evans, *J. Chem. Soc.* p. 1735 (1960); *Proc. Roy. Soc.* **A255**, 55 (1960).

[6] M. Beer, *J. Chem. Phys.* **25**, 745 (1956).

TABLE 19 Phosphorescence of Polyacetylenes[a, b]

Compound	0,0 Band (cm^{-1})	Lifetime (sec.)
$H_3C—(C≡C)_3—CH_3$	22,320	0.5
$H_5C_2—(C≡C)_4—C_2H_5$	18,820	—
Ph—C≡C—Ph	21,870	0.3
$Ph—(C≡C)_2—Ph$	20,270	0.1
$Ph—(C≡C)_3—Ph$	19,380	0.04
$Ph—(C≡C)_4—Ph$	17,150	0.01

[a] All measurements in EPA at −170°C. Ph = Phenyl.
[b] From M. Beer, *J. Chem. Phys.* **25**, 745 (1956).

Similarly various aliphatic aldehydes and ketones have had their phosphorescence properties studied. Formaldehyde[7] shows no phosphorescence. However, at the long-wavelength end of its absorption spectrum a few weak bands can be observed and these have been definitely identified as the singlet–triplet absorption bands. They correspond to a transition that is forbidden both by symmetry and by spin from the ground state to the n,π^* triplet state of the molecule.

A weak phosphorescence is observed for glyoxal[8] (CHOCHO) as well as for deuteroglyoxal (CDOCDO). The 0,0 band of the singlet–triplet transition, which can also be observed in absorption, occurs at 19,544 cm^{-1} (for deuteroglyoxal).

In contrast to this compound, biacetyl[1,9] ($CH_3COCOCH_3$) displays a very intense phosphorescence. It can be observed both in solid and in liquid solution and in the vapor state.

Simple aliphatic ketones such as acetone,[10] methyl ethyl ketone,[10]

[7] J. Brinen and L. Goodman, *J. Chem. Phys.* **35**, 1219 (1961); A. Cohen and C. Reid, *J. Chem. Phys.* **24**, 85 (1956); G. Robinson, *Can. J. Phys.* **34**, 699 (1956); J. C. D. Brand, *J. Chem. Soc.* p. 858 (1956); for comprehensive reviews, see J. W. Sidman, *Chem. Rev.* **58**, 689 (1958) and G. Robinson and V. DiGiorgio, *Can. J. Chem.* **36**, 31 (1958).

[8] J. C. D. Brand, *Trans. Faraday Soc.* **50**, 431 (1954).

[9] J. W. Sidman and D. S. McClure, *J. Am. Chem. Soc.* **77**, 6461 and 6471 (1955); M. Bhaumik and M. El-Sayed, *J. Chem. Phys.* **42**, 787 (1965); H. L. J. Backstrom and K. Sandros, *Acta Chem. Scand.* **14**, 48 (1960).

[10] D. S. McClure, *J. Chem. Phys.* **17**, 905 (1949).

diethyl ketone,[10] cyclopentanone,[11] etc. phosphoresce at 4400–4500 Å. The phosphorescent lifetimes of aliphatic carbonyl compounds are very short; they amount to a few milliseconds or less.

2.7. Alkaloids, Drugs, and Biochemical Systems

Numerous alkaloids, drugs, and biochemical systems that contain aromatic or heterocyclic complexes show phosphorescence at low temperatures. This branch of phosphorescence spectroscopy, in which clinical chemists, biochemists, and biophysicists are especially interested, is acquiring increasing significance at the present time, among other reasons because of the important role that the triplet state is likely to play in many biochemical reactions. Within the space of the present monograph only a few aspects and individual results from this already extensive field can be discussed.

Winefordner and his colleagues[1] have measured the phosphorescence spectra of many alkaloids. These consist of one or two broad bands between ca. 300 and 500 mμ. The phosphorescence decay times for the compounds investigated lie between 0.25 second for morphine and 7.4 seconds for yohimbine hydrochloride. A number of interesting relationships seem to exist between the positions of the phosphorescence transitions of the alkaloids and their constitutions, though they were not discussed in detail. It is surprising to observe, for example, that the phosphorescence bands of the compounds used, codeine, morphine, and thebaine, the aromatic complex of which is a benzene nucleus, occur at rather long wavelengths; these alkaloids show broad bands at 500 mμ.

Again, although pyridine itself displays no phosphorescence, it is found that the simple pyridine derivatives nicotine, nornicotine, and anabasine phosphoresce in an alcoholic solution containing sulfuric acid at 77°K. The broad band of these compounds lies at about 400 mμ and its phosphorescent lifetime is 5–6 seconds.

It is to Winefordner and his collaborators[2] that we owe the employment of phosphorescence spectroscopy in studying pharmaceuticals.

[11] S. R. LaPaglia and B. C. Roquitte, *J. Chem. Phys.* **66**, 1739 (1962).

[1] J. D. Winefordner and H. A. Moye, *Anal. Chim. Acta* **32**, 278 (1965); H. C. Hollifield and J. D. Winefordner, *Talanta* **12**, 860 (1965).

[2] J. D. Winefordner and M. Tin, *Anal. Chim. Acta* **31**, 239 (1964); J. D. Winefordner and H. Latz, *Anal. Chem.* **35**, 1517 (1963).

In a later chapter (see Section 4.6) the important analytical applications of these measurements will be thoroughly discussed. For the present only aspirin (acetylsalicylic acid) and procaine (XCIV) are referred to as examples of drugs that phosphoresce at low temperatures.

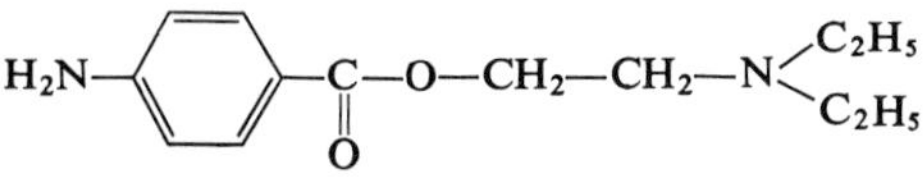

$$H_2N-C_6H_4-\underset{\underset{O}{\|}}{C}-O-CH_2-CH_2-N\begin{matrix}C_2H_5\\C_2H_5\end{matrix}$$

XCIV

Recently the phosphorescence behavior and photochemical properties of nucleic acids have aroused great interest. Bersohn and Isenberg[3] have shown that the phosphorescence of deoxyribonucleic acid (DNA) is a simple superposition of those of its guanine and adenine components. It is interesting that the phosphorescence of DNA is quenched by small amounts of iron(III) or manganese(II) ions. In their electronic properties the dinucleotides are useful models of polynucleotides and Michelson and co-workers[4] have studied the triplet energy transfer from their purine components (adenine) to their pyrimidine components (cytosine) by measurement of their phosphorescence spectra. Haug and Douzou[5] have carried out triplet-ESR research on the nucleic acid building stones in connection with the photochemical properties of these compounds. Among other things it was demonstrated that the photochemical dimerization of orotic acid—of interest in connection with the properties of nucleic acids as carriers of genetic information—proceeds via the triplet state.

Debye and Edwards[6] found that of 18 amino acids investigated only those that contain an aromatic ring, namely tyrosine, tryptophan, and phenylalanine, phosphoresce. The phosphorescence of natural proteins such as albumin, globulin, gelatin, and keratin at low temperatures arises, as the work of various authors shows,[6,7] from these aromatic amino acids, especially tyrosine and tryptophan.

[3] R. Bersohn and J. Isenberg, *Biochem. Biophys. Res. Commun.* **13**, 205 (1963).
[4] C. Helene, P. Douzou, and A. M. Michelson, *Biochim. Biophys. Acta* **109**, 261 (1965).
[5] A. Haug and P. Douzou, *Z. Naturforsch.* **20b**, 509 (1965).
[6] P. Debye and J. D. Edwards, *Science* **116**, 143 (1952).
[7] Yu. A. Vladimirov and Ch'in-Kuoli, *Biofizika* **7**, 270 (1962).

Freed *et al.*[8] have developed a series of new solvents for phosphorescence measurements on proteins, e.g., a mixture of 20% methylhydrazine, 40% methylamine, and 40% trimethylamine. In solvents of this kind these workers have measured the phosphorescence spectra of numerous enzyme proteins such as chymotrypsin, pepsin, and catalase. Here too the phosphorescence clearly originates in the aromatic amino acids tryptophan and tyrosine.

The phosphorescence properties of other biochemically important compounds have also been investigated. Thus Singh and Becker[9] have reported on the phosphorescence of solutions of chlorophylls a and b at 77°K. Both compounds show π–π^* phosphorescence; the bands lie at 8850 and 9250 Å for chlorophyll a and at 8750 and 9150 Å for chlorophyll b.

Similarly Longworth[10] has reported on phosphorescence studies of purines, including purine itself, adenine, guanine, and caffeine, which are also of interest in connection with the phosphorescence of nucleic acids.

In a study of the phosphorescence of various biochemical systems Steele and Szent-Györgyi[11] have obtained information on a hormone, progesterone.

[8] S. Freed, J. H. Turnbull, and W. Salmre, *Nature* **181**, 1731 (1958).

[9] J. S. Singh and R. S. Becker, *J. Am. Chem. Soc.* **82**, 2083 (1960).

[10] J. W. Longworth, *Biochem. J.* **84**, 104 (1962); see also: B. J. Cohen and L. Goodmann, *J. Am. Chem. Soc.* **87**, 5487 (1965).

[11] R. H. Steele and A. Szent-Györgyi, *Proc. Natl. Acad. Sci. U.S.* **43**, 477 (1957).

PART 2 / ANALYTICAL APPLICATIONS OF PHOSPHORESCENCE (SPECTROPHOSPHORIMETRY)

CHAPTER 3 / EXPERIMENTAL PROCEDURES

3.1. Apparatus

Usually the experimental arrangements for measuring luminescence spectra consist of a source of the light used for excitation, a specimen tube containing the luminescent substance, and a device for analyzing and recording the spectrum. Frequently it is advantageous to stimulate the emission not with white light, but with more or less monochromatic radiation. In this case, between the light source and the specimen tube the apparatus includes suitable filters or, better, an (excitation) monochromator that permits the operator to select the wavelength to be introduced. It is almost always essential, when measuring phosphorescence spectra, to have an arrangement for cooling the specimen tube to the temperature of liquid nitrogen or lower. In principle the spectra can be measured without using a phosphoroscope, but they are then recorded simultaneously with the fluorescence spectra that lie at shorter wavelengths. However, the phosphoroscope confers so many advantages in measuring technique that one is compelled to regard it as an essential part of the equipment for determining phosphorescence spectra.

3.1.1. Source of the Exciting Light

The source of the exciting light produces white radiation, both visible and UV; it should be as intense as possible and have an effectively continuous energy distribution. Xenon and mercury vapor lamps are most frequently used. Xenon lamps are inferior to mercury vapor in intensity but give a better continuum. They are preferable for the measurement of phosphorescence excitation spectra (see Section 3.3.2), and are obtainable in a small compact form with ratings between 250 and 500 watts, their introduction into the optical system involving little expense.[1] For the measurement of very weak phosphorescences high pressure mercury tubes are preferable since one can then make use of the high intensities of the principal lines of the spectrum. Data for the dose rate* of such a 500-watt lamp in combination with a large quartz-prism monochromator have been given by Parker[2] for the most important wavelengths.

3.1.2. Filters and Monochromators

Of the two ways in which light of a definite wavelength can be selected from the white light of the source, each has its own advantage. Filters have high transparency and so guarantee more intense excitation, but monochromators produce light of greater spectral purity and permit the introduction of any desired wavelength. The advantages that the use of a monochromator confers may well be decisive.

Whether a prism or a grating monochromator should be used is not at all easy to decide.[1] With the exception of the high frequencies (which, however, are often used for excitation) the resolution is better with a grating monochromator than with a prism instrument of comparable size. The spectrum given by a grating monochromator is linear in the wavelength, an advantage that also applies to the recording of luminescence excitation spectra. When fluorescence spectra are being measured using a grating as the excitation monochromator, interference is frequently caused by scattered exciting light. This does not occur in measuring phosphorescence since scattered light from the source is completely cut off by the phosphoroscope.

The light from the excitation monochromator falls on the specimen tube containing the solution whose luminescence is under investigation. The light emitted then falls onto the entrance slit of either a spectrograph or—for photoelectric recording—a second (the emission) monochromator. In phosphorescence studies, there are, with respect to the direction of excitation and the position of the specimen tube, two particularly important directions of observation. They are shown in Fig. 20. In the "right-angle" method (a) the normals to the entrance and exit slits of the excitation and emission monochromators are

* That is, the intensity of the radiation in microeinsteins per minute.

[1] C. A. Parker and W. T. Rees, *Analyst* **87**, 83 (1962).
[2] C. A. Parker and L. G. Harvey, *Analyst* **86**, 54 (1961).

perpendicular to each other; in the "straight-through" method (b) they coincide. For fluorescence measurements the use of the straight-through method is not usual because of difficulties caused by the exciting light. Phosphorescence studies are more likely to be made with this arrangement since the phosphoroscope eliminates the exciting light quantitatively whatever the geometry of the apparatus; there remains, however, the disadvantage that the "inner filter effects" (see Section

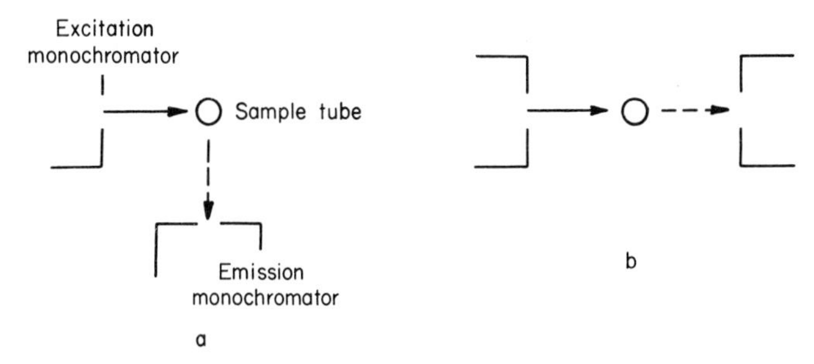

Fig. 20. Positions of specimen tubes in the light paths of phosphorescence spectrometers. (a) Right-angle arrangement; (b) straight-through arrangement.

3.4) appearing with concentrated solutions become particularly serious with the "straight-through" method. Udenfriend[3] has shown that in spectrofluorimetric analyses the most satisfactory shape of the calibration curve is obtained with the "right-angle" method. The same is true for phosphorescence measurements.

3.1.3. The Specimen Tube

For phosphorimetry the specimen tube has a cylindrical shape and must fit into a suitable quartz Dewar flask that holds the refrigerant, usually liquid nitrogen at −196°C. In the simplest arrangement the specimen tubes may be test tubes made of quartz glass. A very convenient arrangement of sample tube and Dewar flask is found in the Aminco-Keirs spectrophosphorimeter discussed below.

3.1.4. The Phosphoroscope

Separation of the rapidly decaying fluorescence from the slowly fading

[3] S. Udenfriend, "Fluorescence Assay in Biology and Medicine." Academic Press, New York, 1962.

phosphorescence, which appears simultaneously, is achieved mechanically by means of a phosphoroscope. As has already been pointed out, this also eliminates the exciting light, so that of the three kinds of radiation that emerge from the specimen tube—fluorescence, phosphorescence, and stimulating radiation—only the phosphorescence succeeds in reaching the spectrograph or the emission monochromator. In the original Becquerel phosphoroscope,[4] the specimen tube is placed between two circular disks that are mounted on a common axis. The disks have openings 1 centimeter in length arranged along the circumference in such a way that apertures in the one disk coincide with obstructions in the other. If the disks are rotated by an electric motor, then the specimen tube "sees" alternately the exit slit of the excitation monochromator and the entrance slit of the emission monochromator or of the spectrograph. That is, in one phase light from the stimulating source falls on the specimen tube, the path from the specimen tube to the receiver being closed, and in the next the phosphorescent light can enter the emission monochromator, while the path from the lamp to the specimen tube is closed. Keirs *et al.* have described a further development of the simple Becquerel phosphoroscope.[5] Lewis and Kasha have used another type.[6] The sample is placed in the middle of a rotating cylinder in which there is an aperture. Then, alternately, light from the source of excitation falls onto the sample and phosphorescent light from the sample onto the entrance slit of the spectrograph. This arrangement has also been used in a modified form in the Aminco-Keirs spectrophosphorimeter (see Section 3.1.6). Parker and Hatchard[7] drive the two disks of the Becquerel phosphoroscope with separate synchronous motors, the geometry and electrical connections being so arranged that without dismantling the apparatus one can choose to measure either the phosphorescence spectrum or *both* phosphorescence and fluorescence spectra.

The phosphorescent light falls onto the entrance slit of either a spectrograph or an emission monochromator, where its spectrum is

[4] E. Becquerel, "La Lumiere, ses causes et ses effets." Gautier-Villars, Paris, 1867; *Am. Chim. Phys.* [2] **27**, 539 (1871). For a mathematical treatment of phosphoroscopes, see T. C. O'Haver and J. D. Winefordner, *Anal. Chem.* **38**, 602 (1966).

[5] R. J. Keirs, R. D. Britt, and W. E. Wentworth, *Anal. Chem.* **29**, 202 (1957).

[6] G. N. Lewis and M. Kasha, *J. Am. Chem. Soc.* **66**, 2100 (1944).

[7] C. A. Parker and C. G. Hatchard, *Trans. Faraday Soc.* **57**, 1894 (1961); *Analyst* **87**, 664 (1962).

resolved, and then is finally recorded by means of either a photographic plate or a photomultiplier. Formerly spectrographs were exclusively employed for phosphorescence measurements. The first workers to employ a monochromator and a photoelectric recorder were Ferguson and his colleagues.[8] Instruments that incorporate these are of great advantage, especially for analytical work in which, among other things, it is important to be able to make measurements rapidly and accurately. Here too grating monochromators will be preferred.

3.1.5. The Detector System

Because in the study of phosphorescence one usually must estimate extremely low intensities, a detector system of the greatest possible sensitivity is necessary.[1] The most important part of this is the photomultiplier, that converts the weak light beam into photoelectrons and multiplies these by a factor of up to 10^8. There are now obtainable 11- to 13-stage tubes with quartz windows and very high sensitivity for a relatively low dark current; Turk[9] has described the various types in a review, though further considerable developments have taken place since that appeared. The output from the photomultiplier is measured by means of a sensitive galvanometer, if necessary after further amplification in order to achieve the highest sensitivity. Parker and Rees[1] have discussed various amplifiers suitable for fluorescence and phosphorescence spectrometers; if their paper is inadequate the appropriate literature on electronics should be consulted. The output from the galvanometer can be linked with a graph plotter (*xy* recorder), thus making possible the rapid and complete presentation of the phosphorescence spectrum. Because the sensitivity of the multiplier is usually greatly dependent on the wavelength, the relative distribution of the intensities of the spectra obtained in this way are considerably distorted from the true quantum spectra though they can be corrected if necessary (see Section 3.3.1).

3.1.6. Phosphorescence Spectrometers

A simple phosphorescence spectrometer with straight-through arrangement and two monochromators is represented schematically in Fig. 21. The lettering has the following significance: a, source of exciting

[8] J. Ferguson, T. Iredale, and J. A. Taylor, *J. Chem. Soc.* p. 3160 (1954).
[9] W. E. Turk, *Photoelec. Spectrometry Group Bull.* **5**, 100 (1952).

light; b, excitation monochromator; c, Becquerel phosphoroscope; d, specimen tube with vacuum vessel; e, emission monochromator; f, photomultiplier; and g, amplifier with curve tracer.

A good phosphorescence spectrometer ought to combine the greatest possible sensitivity with sufficient resolution to permit the measurement of weak emissions. Factors that favor high sensitivity are a powerful source of exciting light, high-intensity monochromators, good geometrical design of the apparatus, and above all a sensitive photomultiplier. Parker and Hatchard [7] have described a particularly efficient,

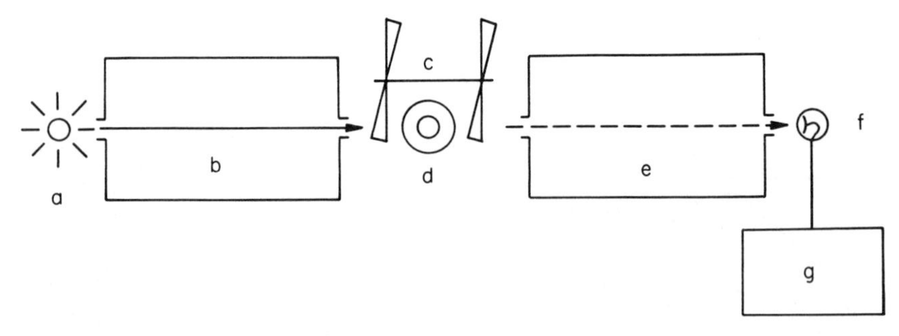

Fig. 21. Diagram of a simple phosphorescence spectrometer with the straight-through arrangement.

very sensitive phosphorescence spectrometer. With this apparatus it is possible to determine phosphorescence to fluorescence ratios down to less than 10^{-5}. The extremely weak phosphorescence that alcoholic solutions of phenanthrene display at room temperature can be measured with it.[10] Muel and Hubert-Habart [11] have reported on a very sensitive arrangement by which weak phosphorescences as far as about 10,000 Å can be measured.

Among the commercially available instruments, several large UV spectrometers, such as the Beckman-Universal, can be used for studying phosphorescence if the necessary supplementary parts are attached. The well-known Aminco-Bowman spectrofluorimeter has been provided with a phosphorescence attachment and has found quite wide distribution as the Aminco-Keirs spectrophosphorimeter.[12] The construction

[10] C. A. Parker, *Chem. Brit.* **2,** 160 (1966) (see Fig. 7).
[11] B. Muel and M. Hubert-Habart, *J. Chim. Phys.* **55,** 377 (1958).
[12] See brochures of the American Instrument Co., Inc., Silver Spring, Maryland.

of this instrument is reproduced schematically in Fig. 22. The light of a xenon lamp (a) is resolved by the grating monochromator (b). The radiation that emerges from this has its wavelength controlled manually within the range 200–800 mμ, and falls on a small quartz specimen tube (c) (shown in section at the right of Fig. 22) in which the solution

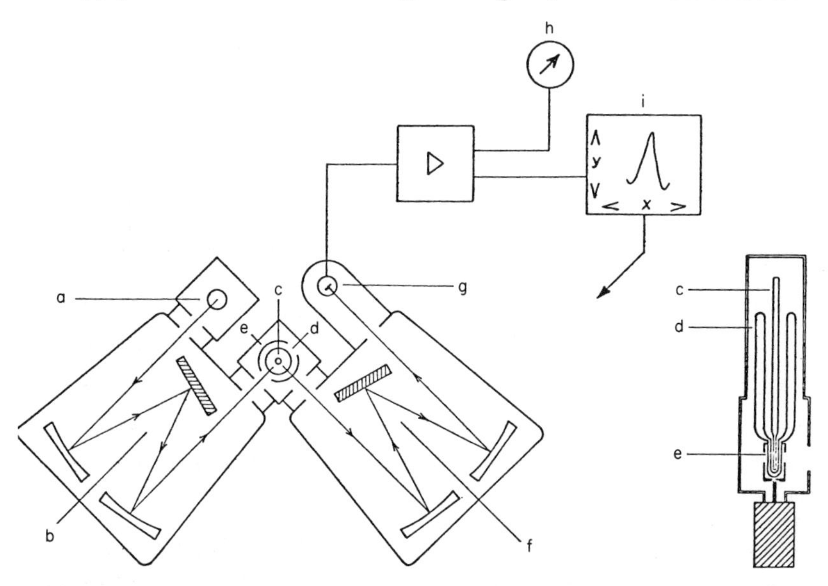

Fig. 22. Schematic representation of the Aminco-Keirs spectrophosphorimeter. [According to M. Zander, *Angew. Chem. Intern. Ed. Engl.* **4**, 930 (1965).]

under investigation is placed in the lower part of an unsilvered quartz vacuum flask (d). The vacuum flask and specimen tube are arranged on the central axis of the phosphoroscope (e). The latter is similar to that used by Lewis and Kasha and consists of a rotating cylinder in which two equally large apertures are found diametrically opposite to each other. The speed of rotation of the cylinder can be varied smoothly. The phosphorescence generated in (c) is then resolved in the grating monochromator (f), which also covers the range 200–800 mμ. The light that leaves (f) falls on a photomultiplier (g), which may be selected by the operator (e.g., RCA 1P28 in the standard model), and is there converted into an electrical signal to be amplified and measured in the photometer (h). This photometer is convenient for use in measuring the key

bands, which can be picked out manually with the monochromator (f). To measure complete phosphorescence spectra, the emission monochromator is electrically operated, and its driving motor is coupled with the photometer output and an *xy* plotter (Electro-Instruments Inc., San Diego 12, California). The sensitivity of the photometer can be varied over a wide range. The resolution of the monochromators (b) and (f) can be adjusted by altering the exit slit of (b) and the entrance and the exit slits of (f) as well as by the choice of the entrance and exit slits on the cuvet housing.

3.1.7. Phosphorescence Decay Phenomena

With arrangements such as those described, it is generally possible also to measure phosphorescence decay phenomena in a simple way. For this purpose the variation of the intensity of a selected band is recorded as a function of the time after interrupting the excitation. A graph plotter cannot be used if the decay takes place too rapidly, i.e., if the mean phosphorescent lifetime is less than about 0.5 second. If the decay is more rapid an oscillograph is necessary and the display on the screen is then conveniently photographed with a Polaroid camera and evaluated. Numerous arrangements for measuring phosphorescence decay processes have been described in the literature,[13–15] that communicated by Kellog and Schwenker[14] seems to be particularly efficient. Van Roggen and Vroom[15] have published a method in which, with the help of a generator with variable constants, an exponential curve is developed and varied continuously until it matches the observed decay curve. The mean phosphorescent lifetime can then be determined from the parameters introduced into the generator.

3.1.8. Polarization and Quantum Yield of Phosphorescence

The relative polarizations of phosphorescence can be measured with the help of such simple additions to the apparatus as are available for, say, the Aminco-Keirs spectrophosphorimeter. Dörr[16] has published a recent review on the spectroscopy of polarized luminescence.

[13] D. S. McClure, *J. Chem. Phys.* **17**, 905 (1949); K. Skarsvåg, *Rev. Sci. Instr.* **26**, 397 (1955); A. Martinez, *Compt. Rend.* **255**, 491 (1962).

[14] R. E. Kellog and R. P. Schwenker, *J. Chem. Phys.* **41**, 2860 (1964).

[15] A. van Roggen and R. A. Vroom, *J. Sci. Instr.* **32**, 180 (1955).

[16] F. Dörr, *Angew. Chem.* **78,** 457 (1966); see also, M. A. El-Sayed and S. Siegel, *J. Chem. Phys.* **44**, 1416 (1966).

McClure and his colleagues[17] have discussed the experimental difficulties of measuring the absolute quantum yields of fluorescence and phosphorescence. Melhuish[18] has described an arrangement which, essentially, combines a rhodamine B quantum counter with a photomultiplier and shows constant sensitivity between 220 and 590 mμ.

3.2. SOLVENTS

The experimental conditions that promote high intensities of phosphorescence are intense stimulation, low temperatures, and rigid media. These conditions will generally be chosen for spectrophosphorimetric analyses.

Since liquid nitrogen (b.p., −196°C) is a very convenient refrigerant and is easily obtainable, substances and mixtures of substances that set either to glassy or crystalline solids at −196°C play an especially important role as solvents in spectrophosphorimetry.

Three groups of rigid media can be distinguished:

(1) organic solvents or mixtures of solvents that set to glasses at −196°C without either crystallizing or breaking up
(2) glassy synthetic materials
(3) solvents that set to crystalline form

Each of these different types of solvent has advantages and disadvantages. Nevertheless, the most important are the organic solvents that form glassy solids. Solutions of this kind can be quickly and easily prepared and, because of their vitreous nature, little reflection occurs, so that few special difficulties are encountered in making quantitative analytical measurements.

By far the most frequently used rigid solvent is that known as EPA, a mixture of ethanol, isopentane, and ether in the proportions by volume of 2:5:5. It sets at the temperature of liquid nitrogen to a rigid, transparent glass that, if the components are dry, shows neither turbidity nor separation of crystalline material and has no tendency to break up.

Obviously it is important that the solvent should itself show the smallest possible phosphorescence. EPA of very good quality and having very little self-phosphorescence is commercially available, e.g., from the

[17] E. H. Gilmore, G. E. Gibson, and D. S. McClure, *J. Chem. Phys.* **20**, 829 (1952).
[18] W. H. Melhuish, *J. Opt. Soc. Am.* **54**, 183 (1964).

American Instrument Co., Inc., Silver Spring, Maryland. The user can, however, prepare the mixture himself. Frequently, this is actually an advantage, as will be shown later. In purifying the solvents it is especially important to remove traces of aromatic and heterocyclic compounds, and for this, adsorption techniques are well suited.[1] In spite of careful purification, however, it is frequently not possible to obtain solvents absolutely devoid of phosphorescence, and this self-phosphorescence can often enable the (lower) limits for the detection of phosphorescent substances to be determined.

TABLE 20 Low-Temperature Glasses for Phosphorimetry (77°K)[a]

Components	Proportion by volume
1. Ethanol:methanol	4:1, 5:1, 5:2, 1:9
2. Isopropyl alcohol:isopentane	3:7
3. Alphanol 79	Commercially obtainable mixture of primary alcohols
4. Butanol:ether	2:5
5. EPA:chloroform	12:1
6. Triethylamine:ether:*n*-pentane	2:5:5
7. Di-*n*-propyl ether:isopentane	3:1
8. 2-Methyltetrahydrofuran	
9. Ethyl iodide:isopentane:ether	1:2:2
10. *n*-Pentane:*n*-heptane	1:1
11. Methylcyclohexane:*n*-pentane	4:1, 3:2

[a] According to F. J. Smith, J. K. Smith, and S. P. McGlynn, *Rev. Sci. Instr.* **33**, 1367 (1962), J. D. Winefordner and P. A. St. John, *Anal. Chem.* **35**, 2211 (1963).

Easily soluble substances can be dissolved directly in ready-mixed EPA at room temperature. The preparation of the solutions for measurement is particularly easy in this case. However, in analytical work, substances are frequently met with which cannot be brought into solution in EPA at room temperature. It is then expedient to prepare a solution of the substance in boiling alcohol and, after cooling, to add the other two ingredients of the EPA, isopentane and ether, in the correct volume proportions.

[1] See, for example, G. Hesse and H. Schildknecht, *Angew. Chem.* **67**, 737 (1955).

If EPA solutions are cooled down to −196°C a contraction of about 30% takes place in the volume. The majority of the concentration data in publications on phosphorescence, like all the author's own measurements recorded in the present monograph, relate to solutions at room temperature.

Beside EPA, there are various other organic solvents and mixtures of these that set to glassy solids at −196°C and which have been suggested and used in phosphorescence spectroscopy. A collection of these is found in Table 20. As well as such polar solvents as EPA, there are also known several nonpolar solvents that are suitable for phosphorescence measurements. Mixtures of methylcyclohexane and isopentane are the most frequently employed of these. Light petroleum (b.p., 58–60°C) also gives a clear, rigid glass.

Several glassy plastics such as poly(methyl methacrylate) (Lucite, Plexiglas) are now gaining significance as solvents in phosphorescence spectroscopy.[2–6] Although the preparation of the solutions is more complicated than when organic solvent mixtures such as EPA are used and although there are usually stronger UV absorption and self-phosphorescence, so that synthetic materials bring additional problems, they do confer advantages for certain special studies. In particular, synthetic materials enable phosphorescence investigations to be carried out over a wide temperature range. They have been used as solvents for investigations into the processes of phosphorescence decay,[3] polarization of phosphorescence,[4] quantum yields,[5] and intermolecular energy transfers.[6] However, particularly in spectrophosphorimetry, they may well have little importance and scarcely displace the organic solvents.

We owe to Shpol'skii[7] the introduction of aliphatic hydrocarbons such as cyclohexane as solvents for spectroscopic investigations at low

[2] E. Laffitte, *Ann. Phys. (Paris)* [12] **10**, 71 (1955); G. Oster, N. Geacintov, and A. Khan, *Nature* **196**, 1089 (1962).

[3] R. Kellogg and R. Schwenker, *J. Chem. Phys.* **41**, 2860 (1964); W. Melhuish and R. Hardwicki, *Trans. Faraday Soc.* **58**, 1908 (1962).

[4] M. El-Sayed, *J. Opt. Soc. Am.* **53**, 797 (1963); *Nature* **197**, 481 (1963).

[5] R. Kellogg and R. Bennett, *J. Chem. Phys.* **41**, 3042 (1964).

[6] R. Bennett, *J. Chem. Phys.* **41**, 3037 (1964); R. Bennett, R. Schwenker, and R. Kellogg, *ibid.* p. 3040; R. Kellogg, *ibid.* p. 3046.

[7] E. V. Shpol'skii, *Usp. Fiz. Nauk.* **71**, 215 (1960); **80**, 255 (1963); *Soviet Phys.—Usp. (English Transl.)* **6**, 411 (1963).

temperatures (see also Section 1.3). Sponer, Kanda, and others[8] have used this kind of solvent for measuring phosphorescence spectra. The snowy, polycrystalline solutions yield spectra that are especially rich in structure and have bands of small half-width. They are therefore well suited for vibrational analyses and also for the identification of substances during analysis. Very insoluble substances do not dissolve sufficiently in those solvents so far known to set to glassy solids. For such compounds, typified by highly condensed aromatic hydrocarbons, 1,2,4-trichlorobenzene has been suggested as a solvent.[9] This sets to crystalline form at low temperatures and the solutions show intense phosphorescence. Trouble arising from the self-phosphorescence of the solvent does not arise if light of wavelength greater than 300 mμ is used for excitation. For aromatic hydrocarbons, the solubility of which permits measurements to be made both in EPA and in trichlorobenzene, it could be shown that, with few exceptions, the spectra in trichlorobenzene are unchanged from those in EPA.

Such highly purified polynuclear aromatic hydrocarbons as phenanthrene that show no measurable phosphorescence in the crystalline state are also suitable for use as matrices. The phosphorescence spectra of aromatic hydrocarbons in a phenanthrene matrix agree[10] with those in EPA apart from a slight broadening of the bands and a red shift of about 250 cm^{-1}.

Although solvents that set to crystalline form offer advantages for certain special purposes such as vibrational analysis and refined identification, and for compounds of low solubility, they are scarcely suitable for quantitative analysis because of the difficulties caused by reflections from the surfaces of the crystals. It may also be mentioned here that the glasses of boric acid and of sugar that were frequently used in older studies of phosphorescence have little value for analytical work. Apart from the fact that it is not easy to prepare the melts, their use entails a series of possible sources of error; these have been discussed by Lower and El-Sayed.[11]

[8] H. Sponer, Y. Kanda, and L. A. Blackwell, *Spectrochim. Acta* **16**, 1135 (1960); Y. Kanda and R. Shimada, *ibid.* **17**, 279 (1961); Y. Kanda, R. Shimada, and Y. Sakai, *ibid.* p. 1; R. Shimada, *ibid.* pp. 14 and 30; Y. Kanda, R. Shimada, K. Hanada, and S. Kajigaeshi, *ibid.* p. 1268.

[9] M. Zander, *Naturwissenschaften* **52**, 559 (1965).

[10] M. Zander, *Z. Elektrochem.* **68**, 301 (1964).

[11] S. K. Lower and M. A. El-Sayed, *Chem. Rev.* **66**, 199 (1966).

Various interesting solvent effects are known in phosphorescence spectroscopy, some of which are important in analytical work. Hammond and his colleagues[12] have compared the positions of the spectra (0,0 bands) of numerous compounds in polar and nonpolar solvents. EPA and a 1:2 ether–ethanol mixture served as polar solvents while 3-methylpentane and mixtures of methylcyclohexane with isopentane were the nonpolar ones. The n–π^* phosphorescences such as are shown by, e.g., acetophenone, benzophenone, anthraquinone, and other carbonyl

TABLE 21 Displacement of Phosphorescence by Solvent[a]

Compound	Phosphorescence 0,0 bands (cm^{-1})	
	Nonpolar solvent	Polar solvent
	n–π^* Phosphorescences	
Benzophenone	24,100	24,350
Anthraquinone	22,000	22,300
Benzil	18,900	20,200
Biacetyl	19,350	20,120
	π–π^* Phosphorescences	
Phenanthrene	21,900	21,750
Naphthalene	21,420	21,480
Carbazole	24,700	24,620
Diphenylene sulfide	24,500	24,400

[a] According to W. G. Herkstroeter, A. A. Lamola, and G. S. Hammond, *J. Am. Chem. Soc.* **86**, 4537 (1964).

compounds definitely undergo hypsochromic displacement when a nonpolar solvent is replaced by a polar one. On the other hand, π–π^* phosphorescences, which are found for hydrocarbons, depend only slightly on the solvent. In the majority of cases a small bathochromic shift is observed in polar as compared with nonpolar solvents. A selection of the results obtained by Hammond and his co-workers[12] has been assembled in Table 21.

[12] W. G. Herkstroeter, A. A. Lamola, and G. S. Hammond, *J. Am. Chem. Soc.* **86**, 4537 (1964).

The phosphorescent lifetimes of a few halogenobenzenes and naphthalenes have been determined by McClure[13] in polar and nonpolar solvents and have been found to be identical within experimental error.

Various authors have shown that, because of heavy atom effects (see Section 1.2), the ratio of the quantum yields of phosphorescence to fluorescence in halogenated solvents (e.g., mixtures of ethanol and ethyl iodide) is greater than in EPA; conversely their phosphorescent lifetimes are shorter. The application of external heavy atom effects in analytical spectrophosphorimetry was first suggested by McGlynn and his co-workers,[14] and Zander[15] has published an exhaustive investigation on this subject. As the perturbing solvent system he used "IEPA," a mixture of 1 volume of methyl iodide and 10 volumes of EPA; this sets to a glassy solid at −196°C. Measurements were made on various polycyclic aromatic hydrocarbons and heterocyclics and their phosphorescence spectra and decay times were determined under strictly comparable conditions of concentration, excitation wavelength, etc. In both solvents the majority of the compounds investigated showed close agreement of the positions of the spectra, the number of bands, and their relative intensity distributions. For an analytically suitable key band—in all cases the bands of greatest intensity before apparatus corrections were applied—the ratio $I_{\mathrm{IEPA}}/I_{\mathrm{EPA}}$ of the phosphorescence intensities (in scale divisions) was found. It was shown that this ratio, which is independent of the conditions of measurement (concentration, excitation wavelength, phosphoroscope speed), varies from substance to substance. Values between 1.5 (1,2:6,7-dibenzpyrene) and 18 (2,3-benzcarbazole) were obtained for the 15 compounds investigated. The fact that the magnitude of this heavy atom effect is specific to the substance suggests a series of possible applications in analysis. For example, for compounds with relatively large $I_{\mathrm{IEPA}}/I_{\mathrm{EPA}}$ ratios, IEPA can be used as solvent instead of the more usual EPA to improve considerably the lower limit of detection. Further details are discussed in another connection in Section 3.3.5.

A medium that is of interest for phosphorescence measurements in another respect is benzophenone. This compound when cooled slowly sets to a crystalline mass, but rapid refrigeration to −196°C produces a

[13] D. S. McClure, *J. Chem. Phys.* **17**, 905 (1949).

[14] S. P. McGlynn, J. Daigre, and F. J. Smith, *J. Chem. Phys.* **39**, 675 (1963).

[15] M. Zander, *Z. Anal. Chem.* **226**, 251 (1967).

clear, glassy, supercooled melt. It can therefore be used in either glassy or crystalline form as a solvent. The phosphorescence that benzophenone shows in both modifications is quenched by the presence of guest molecules that have triplet states lower than that of benzophenone and the phosphorescence of the guest appears simultaneously. It has already been indicated elsewhere that the phosphorescence of many molecules is greater when sensitized by benzophenone than when excited directly (see Section 1.5). Investigations of the phosphorescence of mixed crystals of aromatic hydrocarbons and benzophenone have been carried out by Hochstrasser and Lower[16] (see Section 1.5). Benzophenone may well claim some interest too as a solvent in studying analytical applications of phosphorescence.

3.3. Characterization of Pure Substances by Means of Their Phosphorescence Properties

The phosphorescence of a pure organic compound is characterized by a series of measurable quantities. Of these the following are particularly important, relatively easily accessible, and collectively of considerable significance for planning and carrying out spectrophosphorimetric analyses: phosphorescence spectrum, phosphorescence excitation spectrum, mean phosphorescent lifetime, the ratio ϕ_p/ϕ_f of the quantum yields of phosphorescence and fluorescence, and dependence of the phosphorescence on the solvent.

3.3.1. Phosphorescence Spectrum

To measure the phosphorescence spectrum (with the Aminco-Keirs spectrophosphorimeter (see Section 3.1.6) or some similar instrument) the excitation monochromator is adjusted to a selected wavelength suitable for stimulating the substance under investigation, and the emission monochromator, which is linked to the recorder, sweeps through the wavelength region in which the spectrum is to be found. The compound is in a rigid glassy medium such as EPA and in practice the refrigerant is liquid nitrogen. The speed at which the phosphoroscope rotates is so chosen that when it is raised further no further increase in the intensity of the spectrum can be measured. Obviously if the phosphorescence

[16] R. M. Hochstrasser, *J. Chem. Phys.* **39**, 3153 (1963); R. M. Hochstrasser and S. K. Lower, *ibid.* **40**, 1041 (1964).

decays slowly a lower speed of rotation can be chosen than if it fades rapidly.

In simple apparatus, excitation is with the unfiltered white light of the mercury or xenon lamp. For measuring very weak phosphorescences of reliably pure substances this method is to be recommended, but for spectrophosphorimetric analysis selective excitation offers special advantages that are too important to be surrendered.

The choice of the excitation wavelength—especially for the analysis of mixtures—depends on a series of criteria that have yet to be discussed (see Section 3.5.2). When measuring the phosphorescence spectrum of a sufficiently dilute solution of a single pure compound it is expedient to irradiate with a wavelength that corresponds to a strong absorption maximum of the compound. Suitable wavelengths can therefore be rapidly inferred from the UV spectrum of the substance. In doing this it is necessary to take note of the fact that the majority of the UV spectra to be found in the literature were measured at room temperature, although phosphorescence spectra are obtained at low temperatures. The displacement of the UV spectrum with temperature is, however, generally characteristic of the class of compound. Hence by knowing how these spectra vary with temperature for a few representatives of a particular class the variation of all the rest can be predicted, at least approximately. For the polynuclear aromatic hydrocarbons the α bands (^1L_b) are displaced[1] hypsochromically by ca. 30–50 cm^{-1} on cooling from room temperature to 77°K, while the para (^1L_a) and β bands (^1B_b) are shifted bathochromically about 300 cm^{-1}. Xenon lamps rapidly fall off in intensity in the high-frequency region, so that if a choice has to be made between irradiation in a weak but long-wave absorption band and an intense short-wave band the decision is often in favor of the former.

The phosphorescence spectrum of a pure substance is independent of the irradiating wavelength. If, therefore, different spectra are obtained on exciting with different wavelengths the conclusion may confidently be drawn that the compound is not homogeneous. Each individual compound has been excited in its own absorption region by irradiating with the various wavelengths and the distinct phosphorescence spectrum of each has been produced. Of course it is conceivable that the UV and phosphorescence spectra of the components of a mixture are very

[1] E. Clar, *Spectrochim. Acta* **4**, 116 (1950).

similar. Then the same phosphorescence spectrum will be obtained, within the limits of experimental error, by excitation with different wavelengths. Thus the converse, that when the phosphorescence spectrum of a substance is independent of the wavelength of excitation the substance must definitely be homogeneous, is not true.

In the Aminco-Keirs spectrophosphorimeter, the specimen tube with the rigid solution to be investigated is immersed in the liquid nitrogen that serves as a cooling agent, and so the stimulating light that falls on it and the phosphorescent light appearing from it must pass through the nitrogen. There are other cuvets and cooling devices in which this is not the case. In all kinds of phosphorescence apparatus in which the light passes through the liquid nitrogen it is, of course, decisively important to ensure that the nitrogen contains no snow, i.e., that it is quite dry, and that no gas bubbles rise up in the Dewar flask, since otherwise irregular fluctuations in the intensity of the phosphorescence are recorded, and this makes satisfactory measurement of the spectrum impossible. However, the manipulative techniques for obtaining snow-free and bubble-free nitrogen are easily acquired.

Apart from its dependence on the sizes of the prisms and gratings used in the instrument, the band width depends on the choice of slit widths in the two monochromators. For the purpose of qualitative analysis great importance will be attached to well-resolved spectra, but for quantitative analysis less well-resolved spectra will frequently be tolerated to secure greater sensitivity of the apparatus. Narrow slits are required for well-resolved spectra, wide ones for high sensitivity. A reasonable compromise must be found in each particular case. With the Aminco-Keirs spectrophosphorimeter (using a xenon lamp and photomultiplier RCA 1P28) the following slit widths have been found suitable for qualitative analysis (see Fig. 22): Exit slit *A* of the excitation monochromator b, 3 mm; entrance and exit slits *B*, *C* of the phosphoroscope housing, 0.5 mm; entrance slit *D* of the emission monochromator f, 2 mm; exit slit *E* of the emission monochromator f, 0.5 mm. For quantitative analysis, $A = 3$ mm, $B = 3$ mm, $C = 2$ mm, $D = 3$ mm, $E = 2$ mm.

The phosphorescence spectra the graph plotter traces out do not correspond in the relative intensities of the bands to the true energy spectra (the "quantum spectra"); this occurs particularly because of the dependence of the sensitivity of the photomultiplier on the wavelength.

For theoretical and similar considerations concerning the relationships between constitution and spectra, knowledge of the quantum spectra is a great advantage. For analytical purposes the uncorrected records are more useful since one works with them and correcting them to the quantum spectra introduces new sources of error. In publishing luminescence spectra it is the custom that in predominantly theoretical papers quantum spectra are given and in predominantly analytical papers uncorrected spectra.[2] For the latter the constants of the apparatus are essential and the calibration curve of the photomultiplier is desirable so that readers who are interested in the corrected spectra can themselves make the conversion.

Various methods are known for obtaining the calibration curve of the spectrometer; the most important[3, 4] are measurement with a calibrated tungsten lamp (for the visible region) and the measurement of fluorescence and phosphorescence standards, of which the energy spectra are known. The second method is used particularly often. Suitable fluorescence standards have been suggested by Lippert and his colleagues[4] and by other workers. In the ranges in which the sensitivity of the receiver changes rapidly with the wavelength, the calibrations are unreliable. A detailed description of the procedure with fluorescence standards has been given by Lippert[4] and by Dann and Nickel.[5]

3.3.2. Phosphorescence Excitation Spectrum

To measure a phosphorescence excitation spectrum the emission monochromator is clamped at a wavelength at which an easily measurable phosphorescence band appears, and the excitation monochromator, which is coupled to the curve tracer, sweeps through the range of wavelengths in which the UV spectrum of the substance under investigation is to be found. Thus in the phosphorescence excitation spectrum, one measures the dependence of the phosphorescence intensity on the stimulating wavelength. In the ideal case (the brightness of the lamp and

[2] J. H. Chapman, Th. Förster, G. Kortüm, C. A. Parker, E. Lippert, W. H. Melhuish, and G. Nebbia, *Appl. Spectry.* **17**, 171 (1963).

[3] C. A. Parker and W. T. Rees, *Analyst* **85**, 587 (1960); H. V. Drushel, A. L. Sommers, and R. C. Cox, *Anal. Chem.* **35**, 2166 (1963); C. E. White, M. Ho, and E. Q. Weimer, *ibid.* **32**, 438 (1960).

[4] E. Lippert, W. Nägele, I. Seiboldt-Blankenstein, U. Staiger, and W. Voss, *Z. Anal. Chem.* **170**, 1 (1959).

[5] O. Dann and P. Nickel, *Ann. Chem.* **667**, 101 (1963).

the yield of phosphorescence are independent of the wavelength) the phosphorescence excitation spectrum corresponds accurately with the UV spectrum of the substance. In reality the excitation spectra are distorted by comparison with the absorption spectra, since the dose rates of mercury and xenon lamps vary with the wavelength. Moreover, for a few compounds the phosphorescence quantum yields are functions of the excitation wavelength (see Section 1.4). In order to convert measured into true excitation spectra the energy distribution of the source of excitation must be known. The method by which measured luminescence excitation spectra are corrected has been described in detail by Parker and Rees [3].

True phosphorescence excitation spectra can be very useful in analysis. It is possible to determine the UV absorption of a substance via the excitation spectrum using very much smaller concentrations than are possible in direct measurement of the UV spectrum. With mixtures it is often possible via the excitation spectrum to obtain the UV spectrum of an individual phosphorescent compound, whereas direct measurement of the UV absorption gives the combined curve resulting from the superposition of the absorption spectra of the various components.

Parker [6] has described a measuring device that records directly the true excitation spectrum; it was developed especially for fluorescence measurements but is applicable also to phosphorescence spectra. Such an appliance is extraordinarily useful. One will often, especially in routine investigations, reject the tedious point-by-point conversion of measured into true excitation spectra, which one has to carry out with the equipment that is commercially available at present, and will be satisfied with the measured excitation spectra. These also have a certain analytical value and often, even in mixtures, enable the UV absorption of individual compounds to be recognized. Apart from the circumstances discussed above, phosphorescence excitation spectra can be falsified by a too-high recording speed. The movement of the excitation monochromator must be as slow as possible, especially when the phosphorescence decays slowly.

3.3.3. Mean Phosphorescent Lifetime

According to all previous experience the process of the decay of the phosphorescence takes place according to a strictly exponential law,

[6] C. A. Parker, *Nature* **182**, 1002 (1958).

if the substance is in EPA or a corresponding organic glass at low temperature. To measure the process (e.g., by use of the Aminco-Keirs spectrophosphorimeter or a similar instrument) excitation and emission monochromators are firmly fixed at suitable wavelengths, the excitation is interrupted, and the decay of phosphorescence intensity is recorded either (for slow processes, i.e., long lifetimes) with an *xy* plotter or (for rapid processes, i.e., short lifetimes) by an oscillograph (the picture on the screen is conveniently photographed by means of a Polaroid camera). If, then, the logarithm of the phosphorescence intensity (in scale units) is plotted against the time, a straight line is obtained the gradient of which is given by $2.303\tau_0$, where τ_0 is the mean phosphorescent lifetime (see Section 1.4).

The strictly exponential course of the phosphorescence decay process (over an interval of several times τ_0) is an excellent criterion of purity. If it is observed, then one can conclude that, at the emission wavelength used, only a single compound phosphoresces. If a strictly exponential decay and a constant value of τ_0 are found over the complete range of the phosphorescence spectrum, then we clearly have a single pure compound. The only case in which this evidence can be misleading is when the two or more components of a mixture have exactly the same mean phosphorescent lifetime, but that will seldom be the case. If, conversely, it is established that a measured phosphorescence decay process cannot be represented by the semilogarithmic plot ($\log I_{\text{phos}}/t$) as a straight line, but only as the superposition of two, three, or more straight lines with different gradients, then it can confidently be concluded that, at the selected emission wavelengths, two, three, or more substances with different τ_0 values phosphoresce.

An important exception to the behavior just described is known. A substance in a solvent in which an external heavy atom effect appears does not show exponential decay of phosphorescence. The phenomenon was first observed by McGlynn and his colleagues[7] in rigid solutions of 1-halogenonaphthalenes in propyl halides (Cl, Br, and I) at −190°C. Zander[8] confirmed it for several polynuclear aromatic hydrocarbons and heterocyclics in a mixture of 10 volumes of EPA and 1 of methyl iodide. In these cases the semilogarithmic plot of the decay process

[7] S. P. McGlynn, M. J. Reynolds, G. W. Daigre, and N. D. Christodouleas, *J. Phys. Chem.* **66**, 2499 (1962).

[8] M. Zander, *Z. Anal. Chem.* **226**, 251 (1967).

produced a series of straight lines of different slopes. The decay process is therefore exhibited as the overlapping of several exponential functions with different time constants (τ_0). McGlynn has discussed the origin of this behavior.[7]

Experimental evidence so far suggests that the mean phosphorescent lifetime is a characteristic constant of a substance for a given temperature and solvent, i.e., is independent of the concentration. It can therefore be used, in addition to the spectrum, for the identification of a phosphorescent compound.[9] The extraordinary simplicity and rapidity with which the determination of τ_0 can be carried out encourages its use in qualitative phosphorescence spectroscopic analysis of pure substances and mixtures. This is illustrated elsewhere by a number of examples (see Section 3.5.3).

3.3.4. The Ratios ϕ_p/ϕ_f

From the ratio ϕ_p/ϕ_f of the quantum yields of phosphorescence and fluorescence it is possible to determine whether a compound phosphoresces or fluoresces more strongly. This is of theoretical significance and also of importance in planning phosphorimetric analysis.

To determine ϕ_p/ϕ_f the total luminescence spectrum (fluorescence *plus* phosphorescence) of the substance at low temperature is measured in an organic glass using a fluorescence spectrometer. To do this using the phosphorescence spectrometer described by Parker and Hatchard[10,11] (see Section 3.1.6), the two circular disks of which the phosphoroscope consists are brought "into phase"; the phosphoroscope is removed from the Aminco-Keirs spectrophosphorimeter (which can be done very quickly) and in this way the instrument is converted into a spectrofluorimeter. The measured total luminescence spectrum is converted into the energy spectrum by means of the calibration curve of the instrument and the areas of the shorter-wavelength fluorescence spectrum and of the longer-wavelength phosphorescence spectrum are determined with a planimeter. The ratio of the areas is identical with ϕ_p/ϕ_f.

Parker and Hatchard[11] have measured ϕ_p/ϕ_f values for compounds of various classes. Their results are reproduced in Table 22. It is clear from these that numerous compounds, such as the aromatic carbonyl compounds (see also Section 2.2), can only be detected and estimated by

[9] M. Zander, *Angew. Chem. Intern. Ed. Engl.* **4**, 930 (1965).
[10] C. A. Parker and C. G. Hatchard, *Trans. Faraday Soc.* **57**, 1894 (1961).
[11] C. A. Parker and C. G. Hatchard, *Analyst* **87**, 664 (1962).

phosphorescence and not by fluorescence. For aromatic hydrocarbons in principle both methods can be applied and further criteria still to be discussed are needed to enable one to decide which of the two luminescence methods is the more suitable in a particular case (see Section 3.7).

TABLE 22 ϕ_p/ϕ_f Values for Some Compounds[a]

Compound	ϕ_p/ϕ_f
Benzene	0.89
Phenanthrene	0.80
Biphenyl	1.4
Triphenylene	5.1
Phenol	0.93
Phenyl-2-naphthylamine	1.7
4-Nitro-*N*-ethylaniline	>10
Benzophenone	>10
Anthraquinone	>10

[a] According to C. A. Parker and C. G. Hatchard, *Analyst* **87**, 664 (1962).

3.3.5. The Dependence of Phosphorescence on the Solvent

Among the effects of solvent on phosphorescence that have already been discussed (see Section 3.2), the most important in spectrophosphorimetry is the heavy atom effect shown by solvents that contain halogen atoms. The mixture known as IEPA, which consists of 10 parts by volume of EPA to 1 part of methyl iodide, is a solvent that produces an external heavy atom effect and hardens without fracture to a glass at −196°C; it is therefore also suitable for quantitative spectrophosphorimetry.[8]

It is of great advantage in analysis if a substance can be characterized by *two* spectra. This is possible, e.g., with coronene, triphenylene, 1,2:7,8-dibenzocarbazole (XCV), and brasan (XCVI) if the spectra are measured in EPA and in IEPA. Although most of the compounds investigated in this connection show extensive agreement between their phosphorescence spectra in EPA and IEPA, those of the four compounds named above are characteristically different in the two solvents,[8]

XCV XCVI

and as examples the spectra of triphenylene and brasan are reproduced in Figs. 23 and 24. In this connection see also Section 4.2.

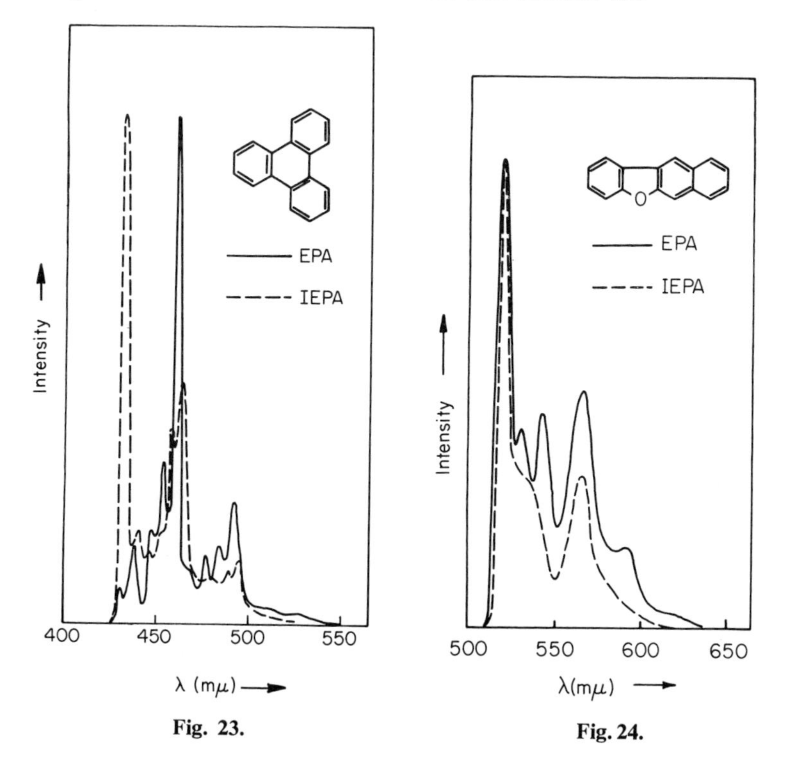

Fig. 23. Phosphorescence spectra of triphenylene in EPA (—) and IEPA (--). The spectra have been normalized so that the most intense band has the same height in each case. [According to M. Zander, *Z. Anal. Chem.* **226**, 251 (1967).]

Fig. 24. Phosphorescence spectra of brasan in EPA (—) and IEPA (--). The spectra have been normalized so that the most intense band has the same height in each case. [According to M. Zander, *Z. Anal. Chem.* **226**, 251 (1967).]

3.4. Foundations of Quantitative Spectrophosphorimetry

The intensity P of the phosphorescence of a solution is proportional to the intensity of the light absorbed by the dissolved substance and to the phosphorescence quantum yield, ϕ_p. If the intensity absorbed is written as the difference between the incident and transmitted intensities I_0 and I_t, then

$$P = \phi_p(I_0 - I_t) \tag{6}$$

By the Lambert-Beer law $I_t = I_0 \cdot 10^{-\epsilon cd}$, where ϵ is the molar extinction coefficient, c the concentration in moles/liter, and d the thickness of the layer of solution in centimeters, and so we obtain from Eq. (6)

$$P = \phi_p I_0(1 - 10^{-\epsilon cd}) \tag{7}$$

Expressing Eq. (7) as a power series we have

$$P = \phi_p I_0(2.3\epsilon cd - (2.3\epsilon cd)^2/2! + \cdots) \tag{8}$$

If the extinction of the solution $E = \epsilon cd$, is less than 0.01, the higher terms of the series can be neglected and Eq. (8) becomes

$$P = 2.3\phi_p I_0 \epsilon cd \tag{9}$$

Equation (9) is the fundamental equation of quantitative spectrophosphorimetry and indicates that in sufficiently dilute solutions a linear relationship exists between the intensity of phosphorescence and the concentration of the solution. It was originally developed for fluorimetry[1] and was carried over to phosphorimetry by Keirs *et al.*[2]

The concentrations at which it is possible to work, and yet to be within the range of validity of Eq. (9), must be decided from case to case. For example, we may consider here the behavior of the aromatic hydrocarbons when studied with the aid of an Aminco-Keirs spectrophosphorimeter.[3] The linear relationship between concentration and intensity of phosphorescence becomes progressively more satisfactory as the extinction of the solution becomes less than 0.01. We may take d to be 0.1 cm, which corresponds roughly to the diameter of the specimen tube of the Aminco-Keirs instrument. On irradiating in the relatively weak α bands (^1L_b) ($\epsilon = 10^2$–10^3) it is found that concentrations $c < 10^{-3}$ M are within

[1] See G. F. Lothian, *J. Sci. Instr.* **18**, 200 (1941).

[2] R. J. Keirs, R. D. Britt, Jr., and W. E. Wentworth, *Anal. Chem.* **29**, 202 (1957).

[3] See M. Zander, *Angew. Chem. Intern. Ed. Engl.* **4**, 930 (1965).

the range of validity of Eq. (9). For irradiation in the intense para and β bands (^1L_a and ^1B_b) ($\epsilon = 10^4$–10^5) greater dilutions ($c < 10^{-5}$ M) are, however, necessary. For example, curves of phosphorescence intensity against concentration have been reproduced in Fig. 25 for triphenylene irradiated in the α and β bands. As can be seen, the concentration range in which linearity prevails is greater when irradiation is in the α band than when it is in the β band.

A quantitative spectrophosphorimetric analysis turns out to be very simple if, in the solution under examination, there are no other substances present apart from the phosphorescent compound. A situation that frequently occurs is one in which a mixture of substances has been separated by paper or thin-layer chromatography and the individual spots have been eluted; spectrophosphorimetry is then utilized solely to determine quantitatively the pure compounds present in the eluates (see Section 3.6). An intense band in the phosphorescence spectrum is chosen as the key band, and for the excitation a wavelength is selected in which the lamp is sufficiently intense and for which the compound under investigation has an intense absorption band. The dilution of the experimental solution is so chosen that it lies within the range of validity of Eq. (9). Next a calibration curve of phosphorescence intensity against concentration is established by means of solutions of known concentration of the pure compound to be estimated. The phosphorescence intensity employed is the uncorrected output of the instrument for the wavelength of the key band. After establishing the calibration curve, the sample being investigated is measured under identical conditions. The concentration of the solution can then be obtained directly from the calibration. Obviously it is important that the brightness of the source should not change [excitation intensity I_0, Eq. (9)] during the measurement of the calibration curve and the test sample.

A rather more complicated situation prevails if the test sample contains other compounds beside the phosphorescent substance to be evaluated and these likewise absorb at the irradiating wavelength. For the present it is quite immaterial whether these accompanying substances themselves phosphoresce or not.

The phosphorescence intensity of a solution is actually, under otherwise identical conditions, dependent not only on the concentration of the dissolved phosphorescent compound but also on the intensity of irradiation. In solutions of a pure compound A the exciting radiation is

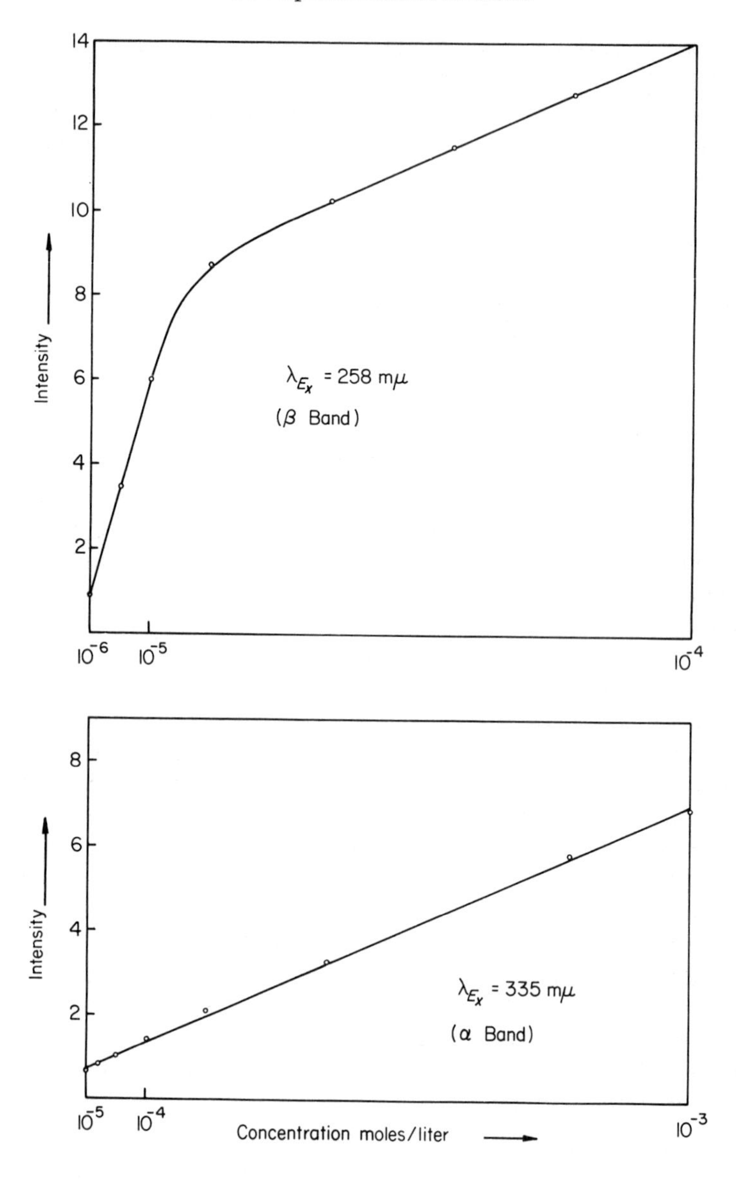

Fig. 25. Phosphorescence calibration curves for triphenylene in EPA at 77°K on excitation in the β band (above) and the α band (below).

absorbed only by this. If there are other absorbing components, B, C, D, etc., in the solution, these take up part of the incident radiation, the whole of which is, therefore, no longer available to excite the phosphorescence of A. Hence, for a given concentration of A, its phosphorescence is smaller in the presence of B, C, D, etc., than in the solution of the pure material. This physically trivial phenomenon, which, however, is of great practical importance in fluorimetry and phosphorimetry, is known as the "inner filter effect."* For a two-component system A, B in which an inner filter effect operates, Eq. (6) takes the form

$$P_A = \phi_{p,A}\left[\frac{\epsilon_A c_A}{\epsilon_A c_A + \epsilon_B c_B}\right](I_0 - I_t) \tag{10}$$

Correspondingly Eq. (8) becomes

$$P_A = \phi_{p,A} I_0 \left(2.3\epsilon_A c_A d - (2.3d)^2 \frac{\epsilon_A c_A(\epsilon_A c_A + \epsilon_B c_B)}{2!} + \cdots\right) \tag{11}$$

Equations (10) and (11) show, as is immediately clear, that an inner filter effect is all the greater, the smaller the ratio of the extinction of component A, which is being measured, to that of the filter component B for the excitation wavelength, and, further, that the effect becomes negligibly small if the total concentration is sufficiently small.[1]

An inner filter effect can cause particularly strong interference if the component A, present in a small amount, must be estimated in the presence of a very large amount of the filter component B. This situation often arises, for example in assaying a phosphorescent impurity A in some very pure product B. It may be assumed that the phosphorescence calibration curve has been determined with solutions that contain only the component A. Because of the filter effect the test sample then shows a lower phosphorescence intensity than the calibration curve does for an equal concentration of A. Thus, in the test sample, the value found for A is wrong, being, in fact, too low.

In principle, there are three possible ways of avoiding or eliminating the inner filter effect:

1. An excitation wavelength is chosen for which the compound to be

* Frequently both the non linearity that is observed between luminescence intensity and concentration in solutions of a pure compound at high concentrations and the reabsorption of luminescence are also known as "inner filter effects." We shall, however, use the term only with the meaning discussed above.

estimated absorbs much more strongly than the accompanying material. However, this possibility cannot often be realized, especially in analyzing complicated mixtures of very similar compounds.

2. The filter components are added to the solutions used for developing the phosphorescence calibration curve, the additions being made in approximately the quantities in which they are present in the test sample. This presupposes that the qualitative composition of the test sample is accurately known and its quantitative composition approximately, and further presumes that all the substances present in the sample are available in pure form. This possibility will only be capable of realization in the analysis of mixtures of relatively simple composition. An example, the spectrophosphorimetric analysis of the impurities in very pure anthracene from coal tar, is discussed in detail elsewhere (see Section 4.1).

3. Work is conducted at dilutions at which inner filter effects are no longer serious. This third way of eliminating inner filter effects is the most generally applicable. It is also successful with mixtures of very complex composition. In practice the procedure is as follows: the assay is carried out for several dilutions, the analytical figures obtained increasing at first with increasing dilution of the experimental solutions. The correct analytical result, no longer falsified by inner filter effects, is recognized by the fact that on further dilution the figures alter no more.

Of course, in principle, it would be possible always to avoid inner filter effects by working from the beginning with very high dilutions. However, high dilutions imply low phosphorescence intensities and necessitate in turn high amplification by the recording instrument. This leads to increased noise in the measuring device and correspondingly reduces the accuracy of the measurement and of the analytical results derived from it. It is therefore more advantageous to estimate the extent of dilution permissible in each particular case and to carry out the analysis for that.

Of the possible sources of error important in the quantitative spectrophosphorimetric analysis of mixtures, the inner filter effect deserves the most attention. By ignoring it, completely false analytical results can be produced.

In fluorimetry the reabsorption of the fluorescent radiation is a frequent source of error. The corresponding phenomenon does not occur in phosphorimetry since phosphorescence spectra lie at appreciably greater wavelengths than UV. Only when colored compounds are present in the mixture can the radiation of the phosphorescent compounds occasionally be absorbed. This case seldom arises. In the same way quenching processes, i.e., processes that reduce the quantum yields, play only a small part in phosphorimetry in contrast with fluorimetry. It was shown earlier that triplet energy transfer in rigid solutions, a process bound up with the quenching of the phosphorescence of one component (the energy donor; see Section 1.5) only appears at very high concentrations; such high concentrations, however, are not generally used in spectrophosphorimetry for the reasons already discussed. The phenomenon of quenching by other substances (e.g., inorganic compounds, phenols, hydrogen sulfide, etc.), that is well known in fluorescence spectroscopy is rare in phosphorescence (but see Sawicki and Pfaff[4]).

The magnitude of the *random* errors to which a phosphorimetric analysis is subject depends, among other things, on the apparatus used. Possible causes of random errors are fluctuations in the brightness of the source lamp, noise in the recording instrument, and variations in the control of the wavelengths. With the Aminco-Keirs spectrophosphorimeter these influences are small and affect the reproducibility of the measurements only to an insignificant extent. The greatest influence on the accuracy of analytical results obtained with the Aminco-Keirs instrument arises from inability to replace the sample tube in exactly the same position in the light path for successive measurements. The relative mean error of phosphorimetric analyses using this spectrophosphorimeter, at the present stage of manipulative skill, amounts to 5–10%.[3] This accuracy is sufficient for many purposes. Details are discussed later with the help of examples (see Sections 4.1–4.7).

Spectrophosphorimetry is a very sensitive method of analysis. The limits of detection for strongly phosphorescent compounds are of the same order of magnitude as for those with strong fluorescence. Thus, for example, triphenylene can be clearly detected[3] even at a dilution of 10^{-9} gm/ml. A thorough comparison of the sensitivity of spectrofluorimetry, spectrophosphorimetry, and UV spectroscopy is given elsewhere (see Section 3.7.3).

[4] E. Sawicki and J. D. Pfaff, *Mikrochim. Acta* p. 322 (1966).

3.5. The Spectrophosphorimetric Analysis of Mixtures

The qualitative and quantitative analysis of a mixture is possible by spectrophosphorimetry if the components present in the mixture differ sufficiently in one or, better, in several of the following measurable properties: phosphorescence spectrum, phosphorescence excitation spectrum, mean phosphorescent lifetime, and sensitivity to external heavy atom effects. Frequently it is the case that only one or a few of the components of a mixture show measurable phosphorescence. These can then usually be very easily detected qualitatively and can also be estimated quantitatively beside the nonphosphorescent components.

3.5.1. Phosphorescence Spectrum

We consider first mixtures the components of which exhibit very similar UV spectra but definitely different phosphorescence spectra. With mixtures of this kind the phosphorescence spectra of all the components present are simultaneously excited by every possible excitation wavelength. The measured spectrum of the mixture is then produced as a superposition of the spectra of the individual components. As a simple example, the spectrum of a mixture of carbazole and phenanthrene is reproduced in Fig. 26. Both compounds can be recognized side by side without any difficulty. The situation does not change—at least qualitatively—if the mixture contains, in addition, nonphosphorescent compounds. The identification of a substance on the basis of its phosphorescence spectrum is carried out by comparison with an authentic spectrum. Naturally spectrophosphorimetry—like every spectroanalytical method—is all the more effective the more authentic spectra of pure substances there are available for purposes of comparison.

For quantitative analysis suitable phosphorescence key bands are chosen for each component of the mixture. The situation is then simple if the phosphorescence spectra of the different components do not overlap at all or do so only slightly. The individual compounds can then be considered independently of each other and calibration curves can be developed for each of them in the way described earlier (see Section 3.4). Hence the amount present in the sample can be determined directly from the intensity of the key band. When partial superposition of the spectra occurs, use can be made, in principle, of the well-known methods for the spectroscopy of polycomponent systems (measurement of several key bands and solution of the system of linear equations). Keirs and co-workers[1] have described this for the phosphorimetry of a two-component system. However, for more complicated mixtures this procedure is difficult to carry out and is subject to large experimental errors. Several techniques of phosphorimetry, which have still to be described, are preferable in such cases.

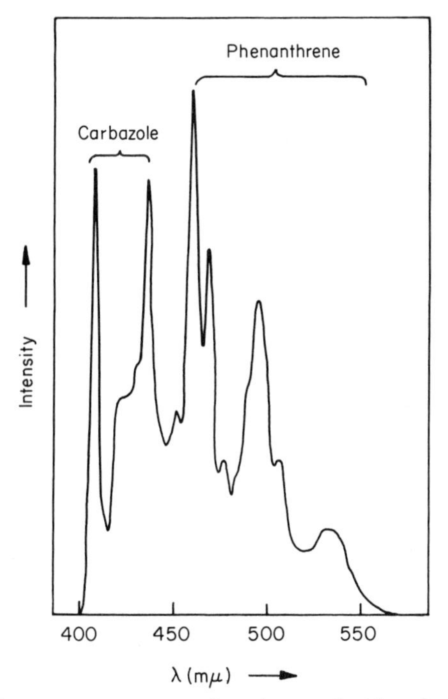

Fig. 26. Phosphorescence spectrum of a mixture of carbazole and phenanthrene in EPA at 77°K on excitation with 290 mμ.

3.5.2. Phosphorescence Excitation Spectrum (*Selective Excitation*)

The spectrophosphorimetric analysis of more complicated mixtures is considerably simplified if the individual components not only have definitely different phosphorescence spectra but also distinctive UV spectra. In such cases, of course, the phosphorescences of the various

[1] R. J. Keirs, R. D. Britt, and W. E. Wentworth, *Anal. Chem.* **29**, 202 (1957).

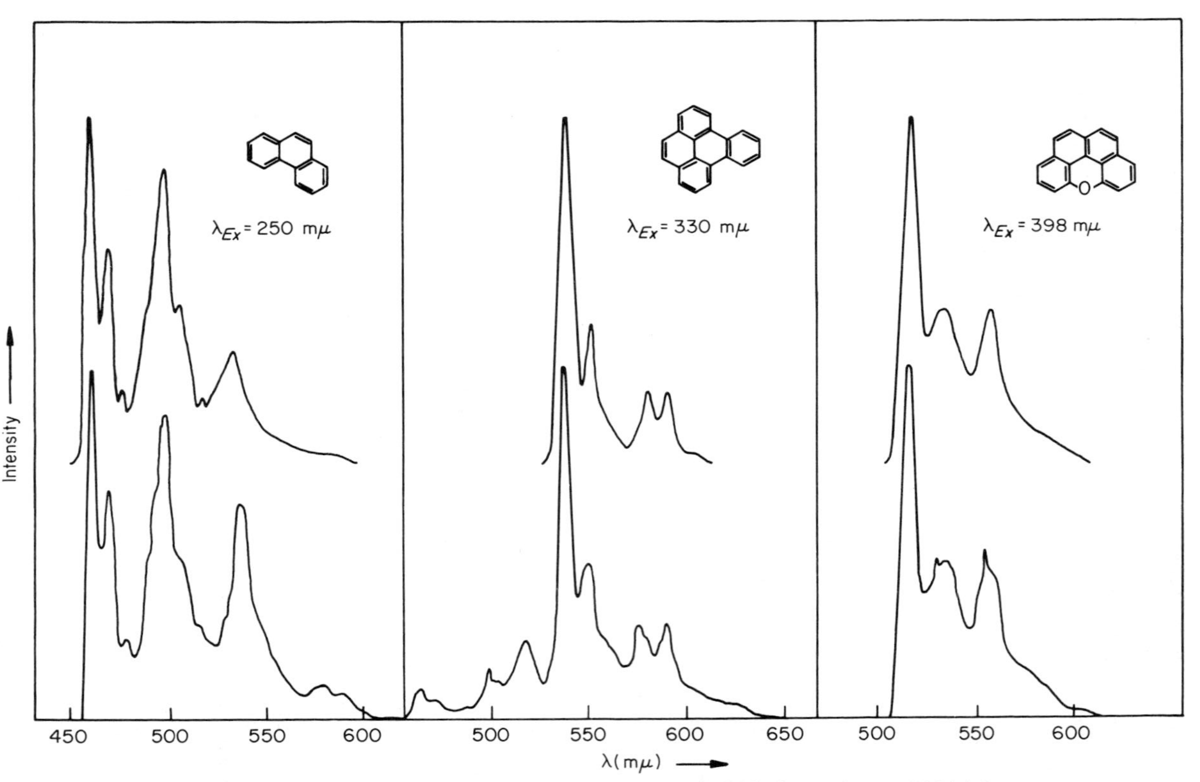

Fig. 27. Phosphorescence spectra of a mixture of 40% phenanthrene, 30% 1,2-benzpyrene, and 30% *peri*-(1,8,9)-naphthoxanthene on excitation with 250, 330, and 398 mμ. The curves displayed above were measured on the pure compounds in each case.

components can be selectively excited. If the phosphorescence of a mixture is excited by a wavelength that is considerably more strongly absorbed by one component than by all the others, then there is obtained predominantly the spectrum of this compound. If, similarly, several excitation wavelengths are employed, it becomes possible to obtain the spectra of the individual components more or less undistorted, provided the UV spectra of the compounds present in the mixture differ sufficiently. This technique, which has long been used in fluorimetry,[2] has been found extremely useful also in phosphorimetry.[3,4]

Figure 27 gives an example. In it are reproduced the phosphorescence spectra of the three-component mixture of phenanthrene (40%), 1,2-benzpyrene (30%), and *peri*-(1,8,9)-naphthoxanthene (30%) excited by three different wavelengths. For each of these wavelengths a different compound gives an intense absorption maximum, and the others give either minima or weak absorption. Above the spectra of the mixture those of the pure components are shown. As can be seen, in each case only one component appears clearly in the spectrum of the mixture. Several practical examples will be discussed later (see Sections 4.1 and 4.2).

3.5.3. Mean Phosphorescent Lifetimes

In addition to spectra, mean phosphorescent lifetimes are of analytical value.[4] Since τ_0 is a constant for the substance if the temperature and solvent are the same (see Section 3.3.3), i.e., it is independent of the concentration of the phosphorescent material and of the possible presence of other compounds, its magnitude can be brought into service for the identification of substances along with their spectra. In this way the reliability of the qualitative evidence in a spectrophosphorimetric analysis can often be considerably enhanced.

Let us suppose that the above three-component mixture is an unknown substance. The analyst obtains, by selective excitation, the spectra reproduced in Fig. 27. Since he did not know beforehand what compounds were present, phosphorescence excitation spectra had to be measured and several excitation wavelengths tested before the optimal results (Fig. 27) could be obtained. The qualitative analysis of the mixture

[2] E. Sawicki, T. R. Hauser, and T. W. Stanley, *Intern. J. Air Pollution* **2**, 253 (1960).
[3] S. P. McGlynn, B. T. Neely, and C. Neely, *Anal. Chim. Acta* **28**, 472 (1963).
[4] M. Zander, *Angew. Chem. Intern. Ed. Engl.* **4**, 930 (1965); *Erdoel Kohle* **19**, 278 (1966).

then proceeds by first comparing the measured spectra with those of known pure substances. For confirmation the mean phosphorescent lifetimes of the three selectively excited phosphorescences of the mixture can be measured and similarly compared with those of the pure compounds. In Table 23 are given the τ_0 values for the three compounds as they were measured in the mixture and in their pure states. It can be

TABLE 23 Phosphorescent Lifetimes of Compounds in a Mixture

	Phosphorescent lifetime (sec.)	
Component	Mixture	Pure compound
1,2-Benzpyrene	2.1	2.0
peri-(1,8,9)-Naphthoxanthene	2.7	2.7
Phenanthrene	3.4	3.3

seen that agreement is within the accuracy of the measurement. Thus the qualitative analytical result derived from the spectra is confirmed by the estimation of the phosphorescent lifetime. Various practical examples of the application of phosphorescent lifetimes in qualitative spectrophosphorimetry are considered elsewhere (Sections 4.1 and 4.2).

For a wavelength for which the phosphorescence spectra of several components of a mixture overlap, the measured phosphorescence decay process is the additive combination of the decay processes of the individual components. The situation is reproduced for a two-component system in Fig. 28, in which the logarithm of the total phosphorescence intensity for a particular wavelength has been plotted against the time. Winefordner[5] has shown how a diagram of this kind can be used for the quantitative analysis of the mixture. If that linear portion of the curve attributable to the more slowly decaying component S is extrapolated to zero time, then its intercept gives the contribution P_S made by S to the total phosphorescence. P_S is directly proportional to the concentration of S in the mixture. If P_S is then subtracted from the measured total phosphorescence (at zero time) the contribution P_F of the faster-decaying

[5] J. D. Winefordner, *in* "Fluorescence and Phosphorescence Analysis" (D. M. Hercules, ed.), p. 179. Wiley (Interscience), New York, 1966.

component is obtained and this, likewise, is proportional to the concentration of F. This technique can also be applied, in analogous fashion, to more complicated mixtures.

Keirs and his colleagues[1] have already described a variation of this procedure. The phosphorescence of a mixture is first measured with very high phosphoroscope speed, so that the spectra of *all* the irradiated substances are recorded. It is subsequently remeasured with a lower

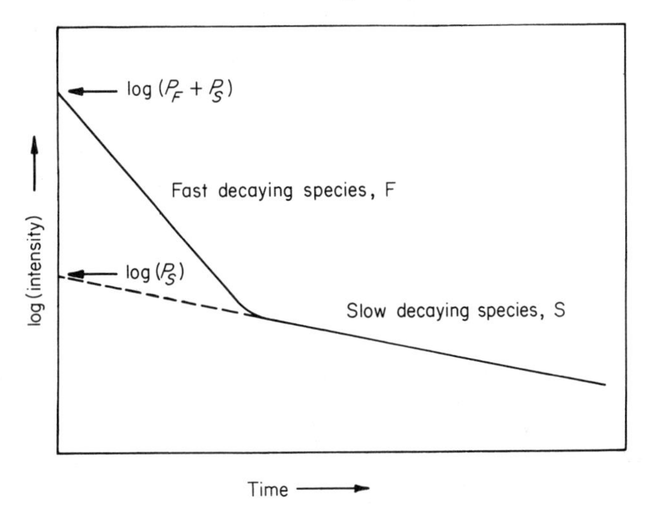

Fig. 28. Phosphorescence decay curves for a two-component system [according to J. D. Winefordner, *in* "Fluorescence and Phosphorescence Analysis" (D. M. Hercules, ed.), p. 180. Wiley (Interscience), New York, 1966].

speed. The rapidly decaying phosphorescence is now completely suppressed or greatly weakened, but the intensity of the slowly decaying phosphorescences remains unaltered. The applicability of the method depends on how much the mean phosphorescent lifetimes of the components present in the mixture differ.

3.5.4. External Heavy Atom Effect

A selective effect that is useful for the qualitative and quantitative spectrophosphorimetric analysis of mixtures, arises from the use of a solvent that shows an external heavy atom effect. Zander[6] has shown

[6] M. Zander, *Z. Anal. Chem.* **226**, 251 (1967).

that a mixture of 10 volumes of EPA and 1 volume of methyl iodide (IEPA, see Section 3.2) increases the phosphorescence intensity of various substances to various degrees. As an example, the phosphorescence spectra of a mixture of 40% *peri*-(1,8,9)-naphthoxanthene (XCVII), 30% picene (XCVIII), and 30% 1,2-benzpyrene (XCIX) in EPA and IEPA have been reproduced in Fig. 29. Both spectra were obtained under exactly the same conditions of excitation wavelength, resolution, etc. In EPA (the continuous curve in Fig. 29) the most intense bands of the

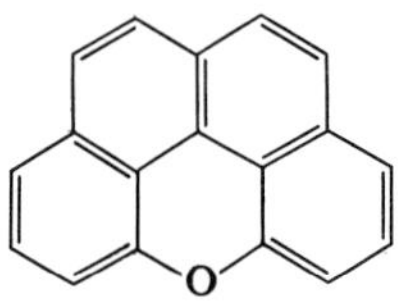

XCVII

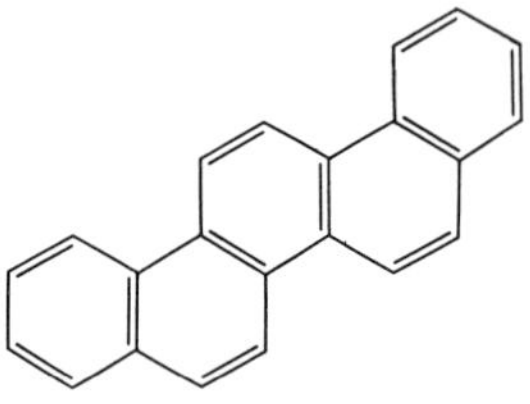

XCVIII

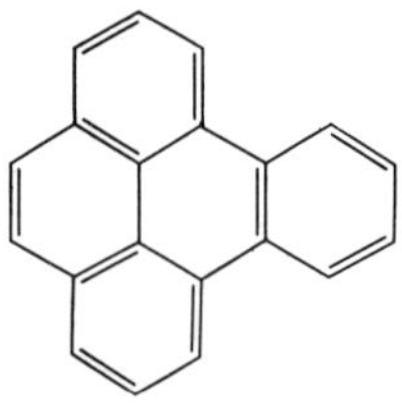

XCIX

mixture arise from the 1,2-benzpyrene (spectrum C, Fig. 29). This compound (key band at 540 $m\mu$) can be determined quantitatively in the mixture in EPA without having to take account of the phosphorescence of the other two components. In contrast *peri*-(1,8,9)-naphthoxanthene shows up relatively weakly in EPA (spectrum B, Fig. 29). To estimate it quantitatively (key band at 515 $m\mu$) in EPA the phosphorescence of picene (spectrum A) must be considered. However, IEPA increases the phosphorescence intensity of the *peri*-(1,8,9)-naphthoxanthene relatively more than those of the other components. In IEPA (the broken curve in Fig. 29) the most intense bands of the mixture result from the *peri*-(1,8,9)-naphthoxanthene. For the determination of this, the phosphorescence of picene can now be neglected, whereby the analysis is simplified and its accuracy increased. Many similar examples have been found. They show that for spectrophosphorimetric analysis of mixtures it is frequently useful to carry out the investigation in both solvents (EPA and IEPA).

For not too complicated mixtures, measurement of the phosphorescence spectra in EPA and IEPA can also frequently establish which bands are related to each other, i.e., are produced by the same compound; they must retain the same relative distribution of intensity on change of solvent.

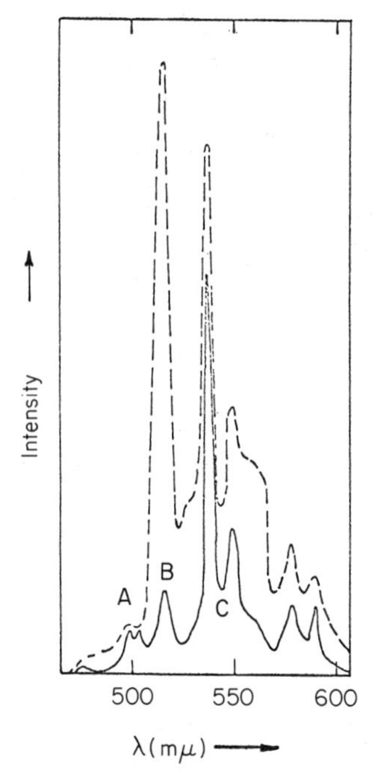

Fig. 29. Phosphorescence spectra of a mixture of 40% *peri*-(1,8,9)-naphthoxanthene, 30% picene, and 30% 1,2-benzpyrene excited by 330 mμ in EPA (—) and IEPA (--). [According to M. Zander, *Z. Anal. Chem.* **226**, 251 (1967).]

3.6. The Combined Application of Spectrophosphorimetry and Chromatographic Methods

Occasionally it is useful, in tackling an analytical problem, to combine spectrophosphorimetry with a suitable chromatographic method. This applies especially to the analysis of mixtures of very complicated composition. The chromatography effects the separation of the mixture into its individual components and the phosphorescence measurement is then used for both the qualitative identification and the quantitative estimation of the separated substances. Investigations into the combination of phosphorescence spectroscopy with paper, thin-layer chromatography, and gas chromatography have all been carried out.

3.6.1. Paper Chromatography

Szent-Györgyi[1] has shown that on paper chromatograms the spots produced by nonfluorescent substances can frequently be rendered visible by their phosphorescence when the paper is cooled in liquid nitrogen (77°K). Here, as in other cases, fluorescence and phosphorescence observations supplement each other since, on the paper, weakly fluorescent compounds generally phosphoresce strongly and vice versa. In contrast with known chemical methods for making visible and locating the spots on paper chromatograms, observation of the luminescence has the advantage that the substances are not altered. The observation can be repeated many times and the eluates of the spots remain available for further investigations.

The method is very simple to carry out. The paper strip is cooled by immersion in a Dewar flask filled with liquid nitrogen, removed, and irradiated briefly with the mercury vapor lamp. When the excitation light has been switched off the phosphorescent spot is easily observed in the darkness and marked with a pencil. By continuous irradiation of the cooled chromatogram the fluorescent spots are also observed. They generally appear more clearly than at room temperature.

The luminescent behavior of the spots supplements the R_f values in a way which is useful for qualitative analysis. Conclusions can be drawn from it with respect to whether a substance appearing in the paper chromatogram fluoresces or phosphoresces more strongly. From the color of the luminescence—especially in the case of phosphorescence—further information can be obtained. The luminescence spectra of the substances adsorbed on the paper can also be measured easily.[2,3] To

[1] A. Szent-Györgyi, *Science* **126**, 751 (1957).

[2] M. Zander and U. Schimpf, *Angew. Chem.* **70**, 503 (1958); M. Zander, *Erdoel Kohle* **15**, 362 (1962).

[3] E. Sawicki and J. D. Pfaff, *Anal. Chim. Acta* **32,** 521 (1965).

determine the phosphorescence spectra the paper must, of course, be at low temperature, and in most cases the spectra are not significantly different from those obtained in solution. Comparison of the measured spectra with the latter can frequently be called upon immediately for purposes of identification. In comparing spectra it must be borne in mind that the spectra measured on the paper generally display rather wider bands and a small red shift compared with those determined in solution.

Frequently the substances that have been separated by paper chromatography and located by luminescence observations will be eluted with a suitable solvent, e.g., ethanol, and the luminescence spectra of the eluates will be estimated. In the case of phosphorescence the methods of investigation discussed earlier (measurement of the spectra with several excitation wavelengths, measurement of the phosphorescent lifetime, quantitative determination with the help of suitable calibration curves, etc.) will be applied to the eluates. In this way the compounds isolated by paper chromatography from a mixture can be quantitatively determined. Because of the great sensitivity of phosphorescence measurements, this technique is also successful if only very small quantities of substance are used in the chromatogram and complicated mixtures are involved.

Zander and Schimpf[2] applied the phosphorescence of the compounds to the paper chromatography of polycyclic aromatic hydrocarbons, the value of the method being established by the analysis of coal tar fractions.

Gordon and South[4] showed that various biochemically important substances such as L-tyrosine, L-phenylalanine, and 2,8-dihydroxy-6-methylaminopurine, which only fluoresce weakly, can be located very simply on paper chromatograms at low temperature because of their intense phosphorescences.

A comprehensive study of the use of phosphorescence in paper chromatography has been published by Sawicki and Johnson.[5] The colors of the luminescence (fluorescence, phosphorescence, or both) of the adsorbed substances at the temperature of liquid nitrogen were reported for some 200 organic compounds of the most varied chemical classes, the greater part of them being of interest in connection with investigations into air pollution. Further, the influence of a basic medium (tetramethylammonium hydroxide in methanol) and of an acid medium

[4] M. P. Gordon and D. South, *J. Chromatog.* **10**, 513 (1963).

[5] E. Sawicki and H. Johnson, *Microchem. J.* **8**, 85 (1964).

(trifluoroacetic acid) on the luminescence behavior of the adsorbed compounds was investigated.

Sawicki and Pfaff[3] have described thoroughly the technique for the determination of phosphorescence spectra directly from the paper chromatograms. The authors gave the phosphorescence detection limits for various compounds such as 1,2-benzpyrene, triphenylene, anthraquinone, *p*-nitroaniline, etc., both in solution in EPA and in the adsorbed state. It was found that a few substances, e.g., 1,2-benzpyrene, can be detected more sensitively in the adsorbed condition, but that the others, e.g., *p*-nitroaniline, are recognized more sensitively in solution.[6]

3.6.2. Thin-Layer Chromatography

The combined application of thin-layer chromatography and phosphorimetry has been described by Winefordner and his colleagues.[7–9] The substances separated from a mixture by this technique are isolated quantitatively by scraping off the individual spots and dissolving them in a suitable solvent. A small volume of the resulting solution is then measured phosphorimetrically in the usual way. The method has been applied, for example, to the determination of nicotine, nornicotine, and anabasine in tobacco[8] and of diphenyl in oranges.[9] The analyses could be carried out quite quickly with substantial accuracy. Thus, the three alkaloids in a tobacco sample were estimated in less than 90 minutes with a maximum standard deviation of 6%.

The direct spectrophosphorimetric measurement of organic compounds on thin-layer chromatograms has proved successful with the method described by Sawicki and Pfaff.[3]

3.6.3. Gas Chromatography

Gas chromatography is the method most frequently applicable to the separation of complicated mixtures. The peaks that appear in the diagrams cannot always be associated with particular compounds on the basis of gas chromatographic data alone. In fact it is frequently necessary

[6] J. D. Pfaff and E. Sawicki, *Chemist-Analyst* **54**, 30 (1965).

[7] J. D. Winefordner, *in* "Fluorescence and Phosphorescence Analysis" (D. M. Hercules, ed.), p. 178. Wiley (Interscience), New York, 1966.

[8] J. D. Winefordner and H. A. Moye, *Anal. Chim. Acta* **32**, 278 (1965).

[9] W. J. McCarthy and J. D. Winefordner, *J. Assoc. Offic. Agr. Chemists* **48**, 915 (1965).

to isolate the compounds separated by gas chromatography in at least sufficient quantity to characterize them by means of suitable spectroscopic methods. Mass spectroscopy and infrared and UV spectroscopy have been used to a great extent for this purpose. Drushel and Sommers[10] have reported on the characterization of gas chromatographic fractions by phosphorescence spectroscopy in a typical case. The basic nitrogen compounds (homologs of pyridine and quinoline, etc.) isolated from a straight-run middle distillate were resolved by gas chromatography and the individual fractions leaving the apparatus were trapped in cooling vessels. Any of these fractions may consist of several substances. The quantities isolated were not sufficient for infrared characterization, but they were quite adequate for UV and phosphorescence spectroscopy. The phosphorescence spectra of the fractions, however, supplied information about their qualitative composition that was not obtainable from the UV spectra since here, as in other cases, the possibility of selective excitation of phosphorescences as well as the characterization of the compounds by their phosphorescent lifetimes was found to be a great advantage.

3.7. Comparison of Spectrophosphorimetry with Other Spectroscopic Methods

The usefulness of a method of analysis may best be judged by comparing it with other related methods. Naturally, in the case of spectrophosphorimetry the comparison that presents itself is with spectrofluorimetry and with UV absorption spectroscopy. The points of view from which the three methods ought to be compared are breadth of application, selectivity, sensitivity, and accuracy.

3.7.1. Breadth of Application

The overwhelming majority of the unsaturated organic compounds show measurable UV absorption, but not all of them reemit the absorbed radiation as measurable fluorescence or phosphorescence. Consequently, the breadth of application of absorption spectroscopy is, in principle, greater than that of the two luminescence spectroscopic methods.

[10] H. V. Drushel and A. L. Sommers, *Anal. Chem.* **38**, 10 (1966).

There are many compounds for which all three methods are suitable for identification and quantitative determination. Neither the experimental difficulty nor the time required is very different for UV, phosphorescence, and fluorescence analyses if a modern recording apparatus is used. The decision with respect to which of the three methods should be preferred for any particular problem can therefore be made solely on the basis of the spectroscopic properties of the system to be investigated.

3.7.2. Selectivity

In luminescence spectroscopy there are available as analytically realizable parameters not only the emission spectra, but also the excitation spectra. This, as well as the fact that not all compounds that absorb measurably also reemit measurably, increases the selectivity of the methods of luminescence spectroscopy, compared with those of absorption spectroscopy. This means, that in complicated mixtures the identification and quantitative determination of individual components is, in many cases, very much simpler by a luminescence spectroscopic method than by UV spectroscopy.

The selectivity of spectrophosphorimetry is greater for several reasons than that of spectrofluorimetry. The compounds whose phosphorescence is measurable are fewer than those for which the fluorescence can be measured. This circumstance limits the range of application of phosphorimetry compared with fluorimetry, but, at the same time, increases its selectivity. The phosphorescence spectra are usually more characteristic than the fluorescence spectra and are therefore frequently better suited for identifying the compounds. This is especially true of the polycyclic aromatic hydrocarbons. As has already been shown, their long phosphorescent lifetimes make many analytical applications possible and further increase the selectivity of the method.

3.7.3. Sensitivity

The sensitivity of luminescence methods is frequently from 10 to 1000 times as great as that of absorption methods. Nevertheless, if the quantum yield of the luminescence to be measured is very small or if the apparatus has a very low sensitivity in the region of the spectrum in which the luminescence appears, the sensitivity of the absorption spectrum can occasionally be greater than that of the luminescence.

Whether a substance can be detected more sensitively, i.e., in greater dilutions, phosphorimetrically or fluorimetrically depends as much on the substance as on the conditions of measurement. We assume first of all that the same conditions of measurement are chosen for the two methods, i.e., the same intensity of the source of the exciting light, the same excitation wavelengths, slit widths, sensitivity of the receiver, etc. The determining factor is then the ratio of the quantum yields of phosphorescence and fluorescence. If the quantum yield of the fluorescence of a substance is definitely greater than that of its phosphorescence then the compound can be detected with greater sensitivity by fluorescence measurements. Conversely if the quantum yield of the phosphorescence is definitely the greater, then the compound is more readily detected phosphorimetrically. In extreme cases generally only one method is applicable. Thus tetracene, for example, shows an intense fluorescence but it has not yet been possible to measure its phosphorescence. Tetracene quinone, on the other hand, displays intense phosphorescence, but fluorescence does not take place at all.

The supposition that we have made above—that the conditions of measurement are the same for both methods—is, however, not usually the case in practice. The phosphoroscope involves, compared with fluorescence determination, a measurable loss of luminescence intensity. On the other hand, in phosphorimetric analyses it is possible to work with considerably wider slits than in fluorimetry because the phosphoroscope completely eliminates scattered (exciting) light. The result is that limits of detection of a compound in phosphorimetry and fluorimetry are still of the same order of magnitude if the quantum yield of phosphorescence is smaller than that of fluorescence by a factor of 10 or 20.

Limits of detection by phosphorescence have been given for a series of compounds. The limits obtained for the same compound by various authors are frequently not exactly comparable with each other since they depend on apparatus parameters such as excitation intensity, receiver sensitivity, noise, and so on, and, in addition, on the background phosphorescence of the solvent and, naturally, also on the excitation wavelength and the key bands that happen to have been selected. (On the difficulty and definition of the term "limit of detection" compare Kaiser.[1])

[1] H. Kaiser, *Z. Anal. Chem.* **209**, 1 (1965).

McGlynn and colleagues[2] have compared the phosphorimetric and fluorimetric limits of detection of durene, naphthalene, and phenanthrene. The fluorescence measurements were carried out both at room temperature and at the temperature of liquid nitrogen (77°K). To round off the investigation, the detection limits of the three compounds by UV spectroscopy were included. These authors found that for aromatic hydrocarbons spectrophosphorimetry and spectrofluorimetry have comparable sensitivity. Sauerland and Zander[3] came

TABLE 24 Limits of Detection of Aromatic Hydrocarbons by Various Processes[a, b]

Hydrocarbon	Limits of detection (gm/ml)		
	Phosphorescence (77°K)	Fluorescence (77°K)	Fluorescence (296°K)
Chrysene	1×10^{-7}	1×10^{-7}	5×10^{-8}
Triphenylene	2×10^{-9}	1×10^{-7}	5×10^{-8}
1,2-Benzpyrene	1×10^{-7}	5×10^{-7}	1×10^{-7}
1,2:5,6-Dibenzanthracene	1×10^{-7}	2×10^{-8}	5×10^{-8}
Coronene	2×10^{-8}	1×10^{-7}	2×10^{-8}

[a] H. D. Sauerland and M. Zander, *Erdoel Kohle* **19**, 502 (1966).

[b] All measurements were made using the Aminco apparatus with a Hanovia mercury-xenon lamp B/1 N.901 and RCA 1P28 photomultiplier. The original publication contains the excitation wavelength, key bands, and a statement of the slit widths. The slit widths were different for the three methods.

to the same conclusion from much more experimental evidence. Some of their measurements are reproduced in Table 24. Wider apertures were chosen for the phosphorimetric determinations than for the fluorimetric. An Aminco apparatus was used for the estimations and, with this particularly narrow slits must be used in low-temperature fluorescence measurements. This is why the fluorimetric limit of detection at 77°K is, in some cases, higher than at room temperature. Triphenylene, for which the ratio of phosphorescence to fluorescence quantum yields is

[2] S. P. McGlynn, B. T. Neely, and C. Neely, *Anal. Chim. Acta* **28**, 472 (1963).

[3] H. D. Sauerland and M. Zander, *Erdoel Kohle* **19**, 502 (1966).

greater than 1, can be detected appreciably more sensitively by phosphorescence than by fluorescence.

It was mentioned earlier that phosphorimetric limits of detection can be improved in some cases if instead of EPA one uses a solvent that shows an external heavy atom effect. As an example, the limits of detection for several compounds in EPA and in a mixture of 10 volumes of EPA with 1 volume of methyl iodide (in this connection see Section 3.2) have been collected in Table 25. The conditions of measurement in both solvents

TABLE 25 Comparison of Phosphorimetric Limits of Detection in EPA and IEPA[a, b]

	Limits of detection (gm/ml)	
Compound	EPA	IEPA
Fluoranthene	2×10^{-6}	5×10^{-7}
Brasan	1×10^{-6}	1×10^{-7}
1,2-Benzcarbazole	2×10^{-6}	1×10^{-7}
2,3-Benzcarbazole	1×10^{-5}	5×10^{-7}

[a] M. Zander, *Z. Anal. Chem.* **226**, 251 (1967).
[b] IEPA = 10 volumes EPA plus 1 volume methyl iodide (see Section 3.2). All measurements were made with an Aminco-Keirs spectrophosphorimeter with Hanovia xenon lamp No. 901 C-1 and RCA 1P28 photomultiplier. The original publication includes the excitation wavelengths and the key bands.

were exactly the same. By the use of the heavy atom solvent instead of EPA, the increase in the sensitivity of detection reached a factor of about 20 in the case of 2,3-benzcarbazole.[4]

Table 26 includes the phosphorimetric limits of detection for a series of polycyclic aromatic hydrocarbons, aromatic carbonyl compounds, nitrogen compounds, and amino acids as well as for a few drugs and alkaloids. The limits are of the order of magnitude of 10^{-7} to 10^{-9} gm/ml.

[4] M. Zander, *Z. Anal. Chem.* **226**, 251 (1967).

TABLE 26 Limits of Detection by Phosphorimetry[a]

Compound	Limit of detection (gm/ml)	Reference[b]
	a. Aromatic Hydrocarbons	
Naphthalene	7×10^{-7}	1
Phenanthrene	3×10^{-8}	2
Chrysene	1×10^{-7}	2
Triphenylene	2×10^{-9}	2
Coronene	2×10^{-8}	2
1,2-Benzpyrene	1×10^{-7}	2
	b. Aromatic Carbonyl Compounds	
Anthraquinone	1×10^{-8}	3
Anthrone	1×10^{-9}	3
Benzil	1×10^{-7}	4
Benzoic acid	5×10^{-9}	5
	c. Aromatic N Compounds	
p-Nitroaniline	1×10^{-8}	3
2-Nitrofluorene	2×10^{-7}	3
2-Aminofluorene	1×10^{-8}	6
	d. Amino Acids	
Tryptophan	2×10^{-9}	6
Tyrosine	1×10^{-8}	6
Phenylalanine	4×10^{-7}	6
	e. Drugs and Alkaloids	
Aspirin	1×10^{-7}	7
Phenacetin	2×10^{-7}	8
Atropine	1×10^{-7}	5
Cocaine hydrochloride	1×10^{-8}	5
Quinine hydrochloride	4×10^{-8}	5

[a] All measurements were made with the Aminco-Keirs spectrophosphorimeter in EPA or ethanol at low temperature. The original publications include, with few exceptions, information concerning the source of the exciting light, type of photomultiplier, slit widths, excitation wavelengths and key bands.

[b] 1. S. P. McGlynn, B. T. Neely, and C. Neely, *Anal. Chim. Acta* **28**, 472 (1963).
2. H. D. Sauerland and M. Zander, *Erdoel Kohle* **19**, 502 (1966).
3. J. D. Pfaff and E. Sawicki, *Chemist-Analyst* **54**, 30 (1965).
4. M. Zander, *Angew. Chem. Intern. Ed. Engl.* **4**, 930 (1965).
5. J. D. Winefordner and M. Tin, *Anal. Chim. Acta* **31**, 239 (1964).
6. See J. D. Winefordner, *in* "Fluorescence and Phosphorescence Analysis" (M. Hercules, ed.), p. 182. Wiley (Interscience), New York, 1966.
7. J. D. Winefordner and H. W. Latz, *Anal. Chem.* **35**, 1517 (1963).
8. H. W. Latz, Ph.D. Thesis, University of Florida (1963).

3.7.4. Accuracy

For several of the methods of spectrophosphorimetric determination described in the literature, relative standard deviations of 2–5% are quoted.[5] In some cases the relative error is greater and can amount to 10%.[6] Clearly the optimum possible accuracy has not yet been reached for phosphorimetric methods of determination. The majority of spectrophosphorimetric methods of analysis described in the literature have been worked out for the Aminco-Keirs spectrophosphorimeter. Improvements appear to be possible both in the apparatus (for which see Section 3.4) and in the technique.

To summarize, it is clear from this comparison of UV spectral analysis, spectrophosphorimetry, and spectrofluorimetry that UV analysis is superior to the two methods of luminescence spectroscopy in its breadth of application, but is inferior to them in selectivity and sensitivity. Phosphorimetry and fluorimetry are nearly comparable methods that are not exclusive, but, rather, in many cases complement each other very satisfactorily. For the quantitative determination of individual compounds in complicated mixtures, phosphorimetry is frequently to be preferred to fluorimetry because of its greater selectivity.

3.8. Sensitized Delayed Fluorescence as an Analytical Method

Triplet energy transfer (see Section 1.5) is quite a general phenomenon. It is observed in gases, in rigid and liquid solutions, and in crystals. In order for a molecule that is in its lowest triplet state to be able to give up its energy to another in its ground state, it is only necessary for the lowest triplet state of the donor to be higher than that of the acceptor. The reaction takes place according to the following scheme:

$$T_D + N_A \rightarrow N_D + T_A \tag{12}$$

where T and N represent the triplet and ground states, respectively, of the donor D and the acceptor A.

In rigid solutions the process is restricted to relatively high concentrations. In liquid solutions at room temperature, triplet energy transfer

[5] J. D. Winefordner and M. Tin, *Anal. Chim. Acta* **32**, 64 (1965); J. D. Winefordner and H. A. Moye, *ibid.* p. 278; M. Zander, *Erdoel Kohle* **19**, 278 (1966); see also J. D. Winefordner, *in* "Fluorescence and Phosphorescence Analysis" (D. M. Hercules, ed.), p. 169. Wiley (Interscience), New York, 1966.

[6] M. Zander, *Angew. Chem. Intern. Ed. Engl.* **4**, 930 (1965).

takes place even at very low concentrations, since the process is then controlled by diffusion. The triplet state of the acceptor, having been excited according to Eq. (12), becomes deactivated in rigid solution both by radiant (sensitized phosphorescence) and by radiationless transitions to the ground state. In liquid solutions no phosphorescence is observed; the process of triplet–triplet annihilation occurs here in competition with the radiationless transition from triplet to ground state (see Section 1.6):

$$T_A + T_A \rightarrow S_A + N_A \tag{13}$$

$$T_A + T_D \rightarrow S_A + N_D \tag{14}$$

In this way one molecule is raised to the singlet excited state S_A, and the other falls to the ground state. From S_A a radiating transition takes place to the ground state and the delayed fluorescence of the acceptor is observed.[1]

Since the processes (12), (13), and (14) take place with great speed, an acceptor can be demonstrated very sensitively, i.e., in very low concentrations, and in the presence of a large quantity of donor. In this, as Parker and his co-workers[2] have shown, lies the analytical importance of the phenomenon of sensitized P-type delayed fluorescence.

The method has been used for the detection of the smallest amounts of impurities in aromatic hydrocarbons. In the simplest case the principal component is itself the donor. On irradiation in an absorption band of this principal component, there are observed the delayed fluorescence spectra of those impurities whose triplet states lie lower than that of the major constituent. In recognizing this, two limitations of the method immediately become apparent. Impurities whose triplet states are higher than that of the main component are not detected in this way. If, on the other hand, many different impurities are present which can behave as acceptors toward the principal constituent, then the sensitized delayed fluorescences of all these impurities appear and the superposition of their spectra can make the identification of the individual components much more difficult.

There is a simple variant of Parker's method by which its breadth of

[1] C. A. Parker and C. G. Hatchard, *Proc. Roy. Soc.* **A269**, 574 (1962); C. A. Parker, *ibid.* **A276**, 125 (1963); C. A. Parker and C. G. Hatchard, *Proc. Chem. Soc.* p. 386 (1962).

[2] C. A. Parker, C. G. Hatchard, and T. A. Joyce, *Analyst* **90**, 1 (1965).

application and its usefulness are greatly increased. By addition of suitable sensitizers (donors) to a mixture, the delayed fluorescences of the individual components can be selectively sensitized. This may be illustrated by an example. The absence of traces of anthracene and 1,2-benzanthracene from a sample of highly purified pyrene is to be tested. Since the UV spectrum of pyrene (the principal component) partly overlaps that of anthracene, direct spectrofluorimetry is scarcely practicable. The use of the pyrene as a sensitizer for the delayed fluorescence of the impurities is likewise inexpedient. Since the triplet states of anthracene and 1,2-benzanthracene are both lower than that of pyrene, the delayed fluorescence of both impurities would be sensitized. Also, since these spectra overlap extensively, it would be difficult to decide with confidence whether in fact the two suspected impurities were present. To test for anthracene there is, therefore, added to the sample a substance the triplet state of which is higher than that of anthracene but lower than that of 1,2-benzanthracene. Acridine orange is such a substance and it sensitizes (λ_{Ex^2} 436 mμ), as would be expected, only the delayed fluorescence of the anthracene and not that of the 1,2-benzanthracene.

The use of this interesting method of analysis is, then, particularly indicated when the triplet states of the components that must be detected are low lying. In these cases phosphorimetry is not very sensitive and may even be quite impracticable. To this extent the two methods supplement each other. It may, however, be mentioned that the technique for measuring delayed fluorescence spectra in liquid solutions is somewhat tedious, in particular because of the necessity of deoxygenating the experimental solutions.

This last difficulty can be avoided if Parker's method is applied to crystals at low temperature instead of to solutions at room temperature. The measurement of the delayed fluorescence of crystallized aromatic hydrocarbons at 77°K is, in many cases, a very sensitive method for the detection of impurities,[3] and is easily carried out. In this, too, the principal component is used as a sensitizer. By measuring the luminescence of the crystal by means of a phosphoroscope, one can observe the delayed fluorescence of those impurities whose triplet states are lower than that of the main component. The sensitized phosphorescence of the impurities often appears beside this. No delayed luminescence of the main component is observed under these conditions.

[3] M. Zander, unpublished data (1966).

CHAPTER 4 / EXAMPLES OF THE APPLICATION OF PHOSPHORIMETRY

4.1. Detection and Determination of Impurities in Polycyclic Aromatic Hydrocarbons

Polycyclic aromatic hydrocarbons play an important role in many fields of research, e.g., in molecular spectroscopy, in semiconductor physics, and in cancer research. In all cases highly purified compounds are needed and knowledge of the impurities is of the greatest importance for planning experiments and discussing results. Similarly, in several technical processes in the chemical industry, the aromatics used as starting materials must be of very high purity.

The majority of the polycyclic aromatics that are commercially available come from coal tar. Since the tar is a very complicated mixture of very similar compounds, the substances isolated still often contain several impurities that are very difficult to remove. The same is true of many of the polycyclic aromatics obtained by synthesis. Thus it is important to have available efficient methods of analysis for the detection and estimation of the impurities in these hydrocarbons. It has been shown that spectrophosphorimetry is a very suitable method of investigation for this purpose. It is, in many cases, more effective than UV and fluorescence spectroscopy, and chromatography, and supplements these methods in many other instances.

An example of the application of spectrophosphorimetry that has been studied with exceptional thoroughness is the investigation of anthracene and its impurities. It has long been known that the principal impurities in the anthracene derived from coal tar are phenanthrene and carbazole. Moreover the presence of sulfur suggests further impurities the nature of which, it is surprising to find, has not yet been completely settled.

The accuracy and sensitivity of the classical chemical methods for the determination of phenanthrene and carbazole in highly purified anthracene—oxidation and treatment with nitrous acid or Kjeldahl analysis—are not sufficient for many purposes. In addition these methods are tedious to carry out.

An elegant gas chromatographic method for the determination of impurities in anthracene described by Sauerland[1] does not achieve the sensitivity of the luminescence spectroscopic methods.

Since the UV spectrum of anthracene overlaps the spectra of phenanthrene and carbazole throughout the whole range of wavelengths, determination of these compounds is difficult if they are present only in small quantities. The limit of detection of phenanthrene in anthracene is of the order of magnitude of 0.1 % in a process described by Schmidt[2] that makes use of statistical methods.

Fluorescence spectroscopy is more sensitive. However, the fluorescence spectra of phenanthrene and carbazole overlap each other extensively, so that fluorimetric determination of the compounds in the presence of each other is difficult.

Carbazole and phenanthrene show intense phosphorescence in the blue and green regions of the spectrum, and anthracene itself phosphoresces very weakly in the red, at 680 mμ. Since the spectra of carbazole and phenanthrene overlap each other only slightly, the prerequisites for the phosphorimetric determination of these principal impurities are ideal. A detailed description of the method of analysis has been given by Zander.[3]

The 0,0 band at 408 mμ is suitable for use as the key band for determining carbazole, but for phenanthrene the band at 500 mμ is used since it is only slightly overlapped by the phosphorescence spectrum of carbazole (in this connection, see Fig. 26). Test analyses showed that, with the concentrations at which these impurities are usually present in anthracene, consideration of the carbazole phosphorescence at 500 mμ is not necessary.

The excitation wavelength for the determination of carbazole and phenanthrene is 290 mμ, at which their extinction coefficients are greater

[1] H. D. Sauerland, *Brennstoff-Chem.* **45**, 55 (1964).
[2] H. Schmidt, *Erdoel Kohle* **19**, 275 (1966).
[3] M. Zander, *Angew. Chem. Intern. Ed. Engl.* **4**, 930 (1965).

than that of anthracene by a factor of about 40. On the other hand, in the samples being investigated, the anthracene has a concentration about 100 times as great, so that the extinctions of the three components are comparable. Consequently a substantial part of the stimulating radiation is absorbed by the anthracene (an inner filter effect; see Section 3.4). It is therefore necessary that the solutions used to set up the calibration curves contain anthracene as well as carbazole and phenanthrene, and indeed in concentrations corresponding approximately to the actual proportions.

TABLE 27 Calibration Mixtures for Spectrophosphorimetric Determination of Phenanthrene and Carbazole in Anthracene[a]

Anthracene, 0.002 *M* (ml)	Phenanthrene, 0.002 *M* (ml)	Carbazole, 0.002 *M* (ml)	Corresponding percentage	
			Phenanthrene	Carbazole
39.92	0.04	0.04	0.1	0.1
19.80	0.10	0.10	0.5	0.5
19.60	0.20	0.20	1.0	1.0
19.40	0.30	0.30	1.5	1.5
19.20	0.40	0.40	2.0	2.0

[a] M. Zander, *Chem.-Ingr.-Tech.* **37**, 1010 (1965).

For the analysis there were first prepared, from 0.002 *M* standard solutions of anthracene, phenanthrene, and carbazole in EPA, the calibration mixtures specified in Table 27. Then, on excitation of the phosphorescence with 290 mμ, the intensities of the key bands were measured. Proceeding as described earlier, these intensity values were plotted against the concentration (as a matter of convenience, immediately in percentage of the component) to form calibration curves. Figure 30 shows the curves for phenanthrene and carbazole in anthracene over the range 0.1–2%. Then, in a similar way, using a 0.002 *M* solution of the test material in EPA, the phosphorescence intensities of the key bands were again measured and the corresponding percentage contents of phenanthrene and carbazole were deduced from the calibration curves.

In highly refined anthracene from coal tar, the impurities are less than 0.1 %. Carbazole and phenanthrene in anthracene can still be determined down to about 0.005 % in the manner described above. For this it is necessary to set up calibration curves for the range 0.1–0.005 %. The mixtures required are conveniently prepared from a 0.002 *M* solution

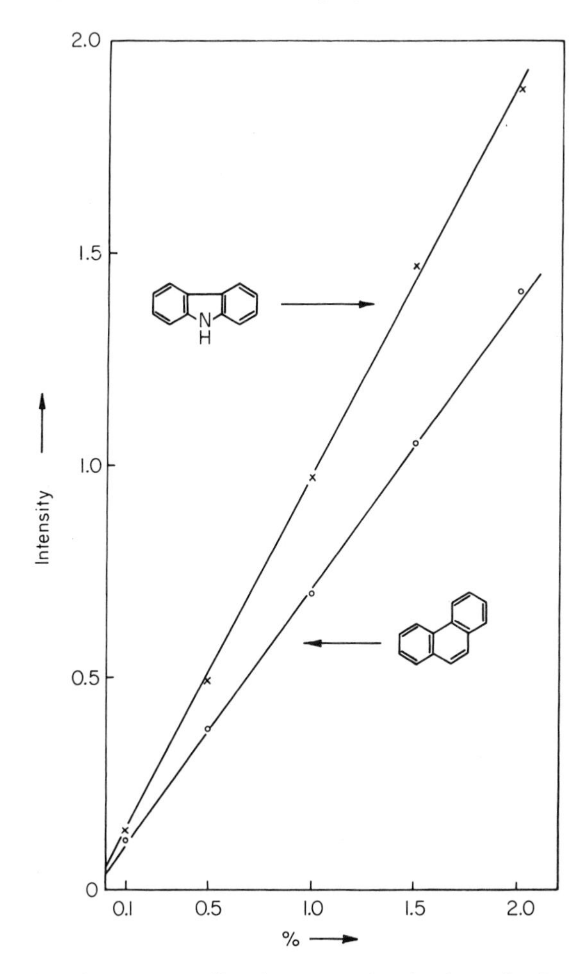

Fig. 30. Phosphorescence calibration curves for the determination of carbazole and phenanthrene in anthracene. [According to M. Zander, *Angew. Chem. Intern. Ed. Engl.* **4**, 930 (1965).]

of anthracene and 0.00002 *M* solutions of carbazole and phenanthrene. Obviously the carbazole and phenanthrene content of the anthracene used for the calibration curves must be below the limits of detection.

The method described was worked out for an Aminco-Keirs spectrophosphorimeter (with mercury–xenon lamp and RCA photomultiplier 1P28). The combination of slit widths quoted earlier for quantitative analyses (see Section 3.3.1) was used. The procedure can, of course, be carried out with other apparatus of comparable sensitivity.

TABLE 28 Testing the Spectrophosphorimetric Determination of Carbazole and Phenanthrene in Anthracene[a]

Anthracene given (%)	Carbazole (%)			Phenanthrene (%)		
	Given	Found	Δ	Given	Found	Δ
99.60	0.20	0.23	0.03	0.20	0.23	0.03
99.00	0.50	0.56	0.06	0.50	0.53	0.03
99.00	1.00	0.90	0.10	—	—	—
99.00	—	—	—	1.00	1.13	0.13
98.80	1.00	0.93	0.07	0.20	0.25	0.05
98.50	1.00	1.05	0.05	0.50	0.67	0.17
98.50	0.50	0.45	0.05	1.00	0.90	0.10
98.00	1.00	1.00	0.00	1.00	0.93	0.07
97.50	2.00	2.10	0.10	0.50	0.57	0.07
97.00	1.00	1.05	0.05	2.00	2.00	0.00
96.50	0.50	0.47	0.03	3.00	2.83	0.17
96.00	3.00	2.95	0.05	1.00	1.17	0.17

[a] M. Zander, *Angew. Chem. Intern. Ed. Engl.* **4**, 930 (1965).

Table 28 gives the results of a series of test analyses. The mixtures investigated were prepared from the purest anthracene, carbazole, and phenanthrene. The anthracene content of the mixtures was chosen between 99.6 and 96%. To establish the reproducibility of the method, several samples of industrial anthracene, as they were obtained from coal tar distillation, were investigated, altogether 6 times, for phenanthrene and carbazole. The values found, the average values, and the deviations are presented in Table 29.

TABLE 29 Testing the Spectrophosphorimetric Determination of Carbazole and Phenanthrene in Anthracene[a]

	Carbazole (%)		Phenanthrene (%)	
	Found	Δ	Found	Δ
Test 1				
	0.59	0.00	0.90	0.02
	0.59	0.00	0.90	0.02
	0.56	0.03	0.86	0.02
	0.59	0.00	0.83	0.05
	0.63	0.04	0.90	0.02
	0.59	0.00	0.86	0.02
Mean	0.59		0.88	
Test 2				
	0.55	0.02	0.70	0.00
	0.49	0.04	0.72	0.02
	0.54	0.01	0.72	0.02
	0.54	0.01	0.67	0.03
	0.52	0.01	0.70	0.00
	0.56	0.03	0.67	0.03
Mean	0.53		0.70	
Test 3				
	0.52	0.01	0.63	0.01
	0.52	0.01	0.64	0.00
	0.49	0.04	0.64	0.00
	0.54	0.01	0.63	0.01
	0.56	0.03	0.67	0.03
	0.54	0.01	0.63	0.01
Mean	0.53		0.64	

[a] M. Zander, *Angew. Chem. Intern. Ed. Engl.* **4**, 930 (1965).

The mean relative error of the method depends, as would be expected, on the magnitude of the concentrations to be determined. For 0.1% of carbazole or phenanthrene, it amounts to ca. 10%; for

0.005%, however, ca. 50%. Nevertheless this accuracy is almost always sufficient.

It has already been mentioned that anthracene derived from coal tar, beside carbazole and phenanthrene, contains sulfurous impurities. The S content of a representative tar anthracene (anthracene content >98%) is 0.06%. It is usually assumed that the S-containing impurity is diphenylene sulfide (C). From the S content a percentage of diphenylene sulfide of 0.35% is calculated. In fact phosphorescence investigations of several tar anthracene samples showed that the diphenylene sulfide content is certainly less than 0.15%. From this it must be concluded that the major part of the sulfur is present in some other form.[4] For several reasons, 5,6-benzothionaphthene (CI) has been suggested and Kruber[5] has detected this compound in coal tar.

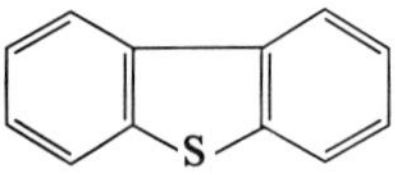

C

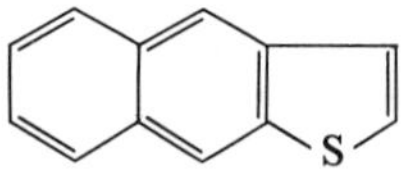

CI

Frequently, in a polycyclic aromatic hydrocarbon, not only the amount but also the nature of the impurity is unknown. Spectrophosphorimetry may also be put to useful service in the identification of the impurities. The technique described earlier is used (selective excitation, identification by spectra and lifetimes, heavy atom effects, etc.). The procedure may be described in detail with pyrene and fluorene as examples.

The spectrophosphorimetric investigation of pyrene (CII) from coal tar (pyrene content ca. 95%) has been described by Zander.[3] In the UV

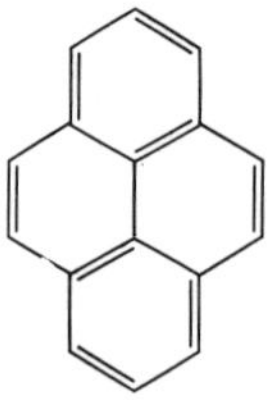

CII

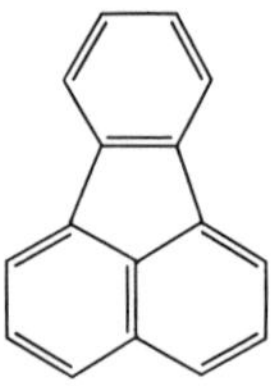

CIII

[4] M. Zander, unpublished data (1965).

[5] O. Kruber and L. Rappen, *FIAT Rev. Ger. Sci.* **36**, 292 (1946).

spectrum an extra band at 288 mμ is observed in addition to the pyrene bands; this could arise from fluoranthene (CIII). The measurement of the phosphorescence by irradiation in a strong absorption band of fluoranthene (360 mμ) gave the phosphorescence spectrum of this compound almost without interference. In Fig. 31 the phosphorescence

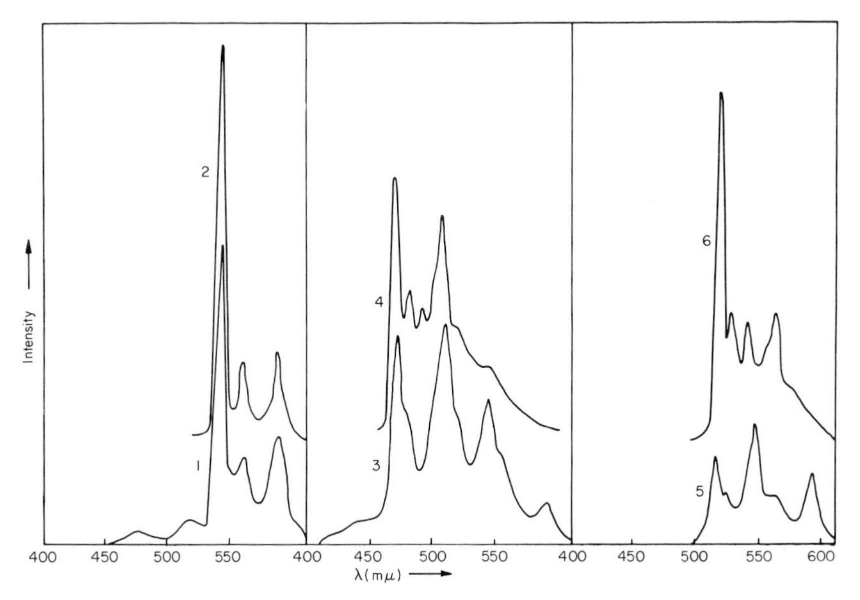

Fig. 31. Phosphorescence spectra of technical pyrene on excitation with various wavelengths. [According to M. Zander, *Angew. Chem. Intern. Ed. Engl.* **4**, 930 (1965).]

spectrum of the pyrene impurity (curve 1) is reproduced beside that of pure fluoranthene (curve 2). For confirmation, the mean phosphorescent lifetimes of the substance accompanying the pyrene and that of pure fluoranthene were both measured for the band at 545 mμ. They agree within experimental error (0.8 and 0.9 second, respectively). Measurement of technical pyrene by irradiation with 290 mμ gave spectrum 3, which belongs to a further impurity. Comparison with the phosphorescence spectra of several components of tar that boil in the same range as pyrene made it seem very probable that the impurity involved here was 1,2-benzodiphenylene oxide (CIV) (phosphorescence spectrum 4 of Fig. 31). As a further check, here too, the phosphorescent lifetimes of the pyrene satellite and of pure 1,2-benzodiphenylene oxide were estimated; within the limits of accuracy of the measurement they agreed (3.9 and 4.0 seconds, respectively). Finally, by exciting the technical pyrene with 340 mμ, a third impurity (curve 5) could be obtained. Since this spectrum overlaps that of fluoranthene on the long-wave side, only the first two bands can be used in the identification. Hence it is probable that this impurity is 2,3-benzodiphenylene oxide (brasan) (CV).

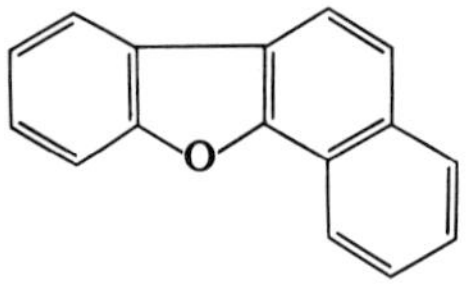

CIV 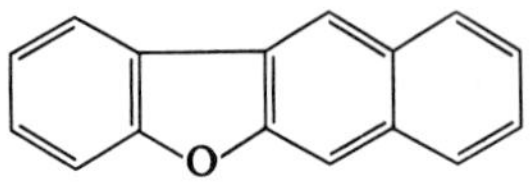 CV

The spectrum of the pure substance is reproduced in Fig. 31 (curve 6) for comparison. To summarize, it is evident that phosphorescence spectroscopic investigation of technical pyrene (CII) permits the definite identification of two impurities, fluoranthene (CIII) and 1,2-benzodiphenylene oxide (CIV), and makes probable that of a third, brasan (CV). In contrast, UV spectroscopy furnishes only a weak hint of one of them (fluoranthene).

Parker[6] spectrophosphorimetrically investigated the impurities in commercially obtainable fluorene and in a substance purified by zone melting. On excitation with 313 mμ, a wavelength that is not absorbed by the fluorene itself, two groups of bands were obtained and clearly arose from two impurities. The group having the shorter wavelength could unequivocally be assigned to carbazole, as the result of comparison with an authentic spectrum. Quantitative spectrophosphorimetric investigation gave 30 ppm as its concentration in the unpurified sample. After zone melting it was less than 2 ppm. The second spectrum could not be assigned beyond doubt, but it seemed very probable that an alkyl derivative of naphthalene was involved. Kanda and colleagues[7] had previously established that commercial fluorene displayed additional phosphorescence bands that were missing from synthetic fluorene. They had attributed these to diphenylene oxide (CVI). It is quite plausible, considering the boiling point range of the coal tar fraction from

[6] C. A. Parker, *Proc. SAC Conf., Nottingham*, 1965 p. 208.

[7] Y. Kanda, R. Shimada, K. Hanada, and S. Kajigaeshi, *Spectrochim. Acta* **17**, 1268 (1961).

which it is obtained, that fluorene contains carbazole or diphenylene oxide as an impurity.

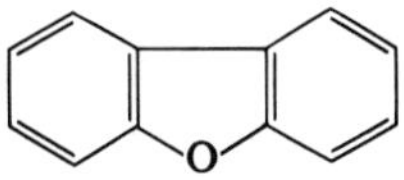

CVI

In many cases it could be shown that the phosphorescence spectrum of aromatic hydrocarbons revealed the presence of impurities that were quite unrecognizable by UV and not definitely recognizable by fluorescence. Traces of impurities can be detected in 3,4-benzpyrene (CVII);

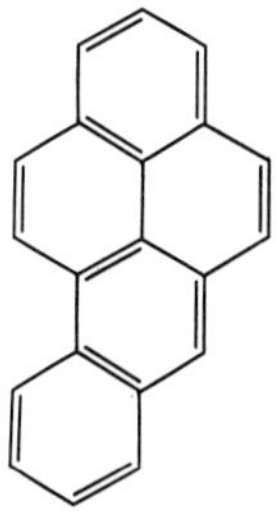

CVII 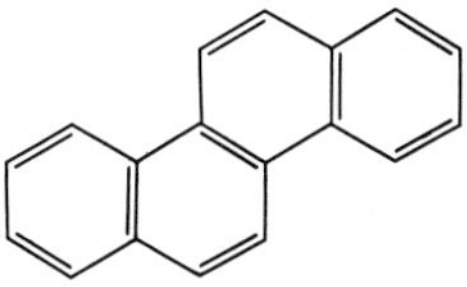 **CVIII**

Parker[6] identified chrysene (CVIII) and demonstrated the presence of another impurity, the identity of which could not be established. The behavior of both impurities was followed up by phosphorescence spectroscopy after zone melting of the 3,4-benzpyrene.

In 1,2:5,6-dibenzanthracene (CIX) obtained from coal tar, picene (CX) could be detected in small quantities. Purification of the dibenzanthracene by means of its addition product with maleic anhydride yielded a product free from picene.[8]

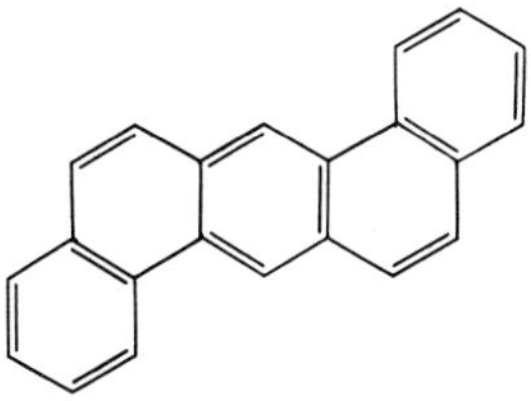

CIX 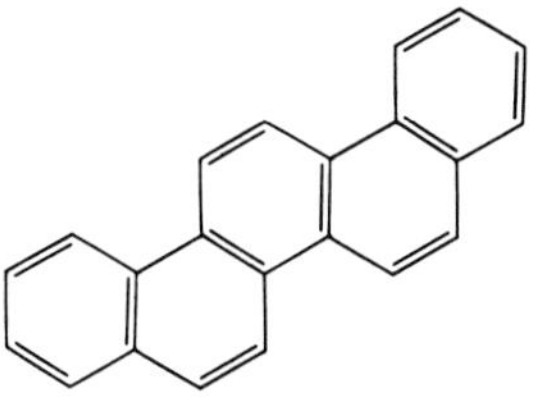 **CX**

[8] M. Zander, *Chem. Ber.* **92**, 2749 (1959).

In 1,2-benzanthracene (CXI) from coal tar chrysene (CVIII) is detectable, and in acenaphthylene (CXII), acenaphthene (CXIII).[4]

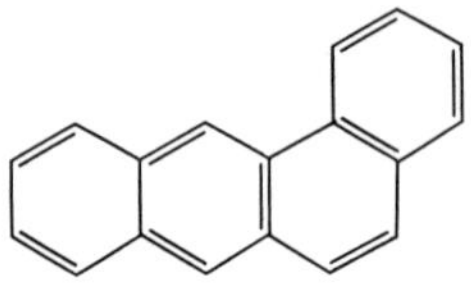

CXI

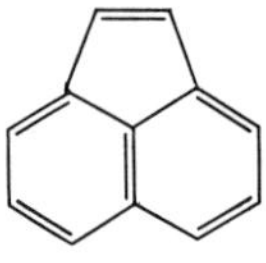

H_2 H_2

CXII 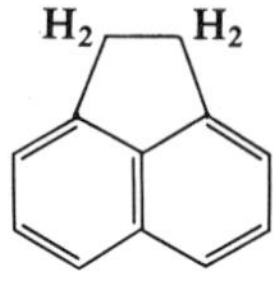 CXIII

In all cases the detection of phosphorescent impurities is very sensitive. Occasionally the detection sensitivity is greatly increased by use of a solvent that shows an external heavy atom effect (see Section 3.2). This was indicated, for example, in the detection of fluoranthene in pyrene.[9] Because the external heavy atom effect is quite specific to the substance (see Section 3.2) it frequently permits a decision to be made with respect to whether the impurity bands result from one or several compounds.

Many polycyclic aromatic hydrocarbons, e.g., the higher annellated acenes such as tetracene and pentacene, are very easily photooxidized. Thus small quantities of quinones are often formed in such purification processes as chromatography on alumina and recrystallization and are very difficult to recognize by ordinary analytical methods. However, most of these quinones phosphoresce very intensely and so can be detected very sensitively in higher annellated hydrocarbons. Among other examples, this has been illustrated[10] by the tetracene-5,12-quinone (CXIV), that is present in small amount in tetracene purified by conventional methods.

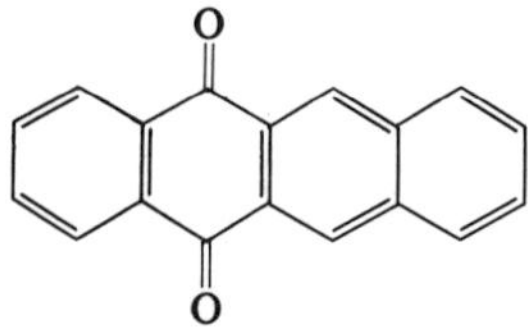

O

O

CXIV

[9] M. Zander, *Z. Anal. Chem.* **226**, 251 (1967).

[10] E. Clar and M. Zander, *Chem. Ber.* **89**, 749 (1956); *J. Chem. Phys.* **43**, 3422 (1965).

4.2. The Analysis of Coal Tar Fractions[1]

At the present time the tar processing industry is much concerned with using its raw material as thoroughly and productively as possible. This demands, among other things, a considerable knowledge of the qualitative and quantitative composition of the fractions resulting from the tar distillation. Even scientists engaged in fundamental research may well be interested in the analytical composition of the tar fractions; for example, those biochemists and medical scientists who are working in the field of cancer research.

Because of the extraordinarily complicated nature of coal tar, even fractions of very narrow range still include a large number of compounds. It is therefore useful to have available the greatest possible number of different physical and chemical techniques that are mutually complementary in range of application. Gas and paper chromatography and UV and flu-

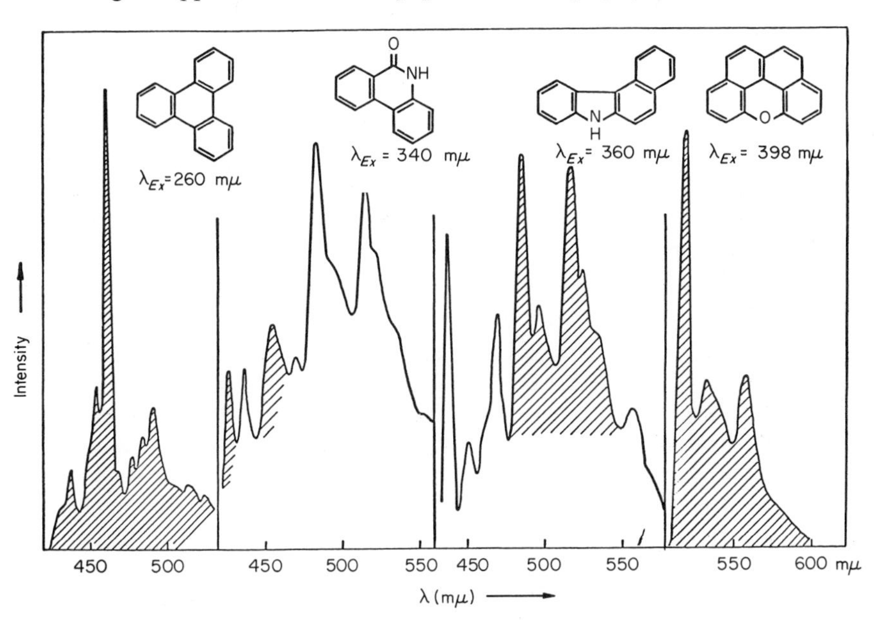

Fig. 32. Phosphorescence spectra of a coal tar fraction boiling at ca. 460°C and 760 torr on excitation with 260, 340, 360, and 398 mμ. [According to M. Zander, *Erdoel Kohle* **19**, 278 (1966).]

[1] M. Zander, *Erdoel Kohle* **19**, 278 (1966).

orescence spectroscopy long ago found their way into the analysis of coal tars. It has been found that spectrophosphorimetry supplements these methods in quite a remarkable way, and that in many cases it is able to supply analytical information that is not accessible by the other methods.

For mixtures that are as complicated as even very narrowly cut coal tar fractions, the method of selective excitation of phosphorescence is particularly important. In Fig. 32 the phosphorescence spectra (in EPA at 77°K) of a coal tar fraction of boiling point ca. 460°C at 760 Torr are reproduced, excitation having been stimulated by different wavelengths. Most of the individual spectra can be assigned to known substances. The range in which each spectrum of the tar fraction agrees with that of the pure compound is indicated on the graphs by the shading. Here, as in other cases, the identification of the substances followed from their spectra and was confirmed by measuring their phosphorescent lifetimes (in this connection see Section 3.5.3).

It was shown earlier that some substances possess different phosphorescence spectra in EPA and in a mixture of EPA and methyl iodide (IEPA, see Section 3.2). This effect occasionally proves useful for the identification of individual compounds in tar fractions of very complicated composition. In Fig. 33 a section of the phosphorescence spectrum of such a fraction is reproduced. The EPA spectrum (the continuous curve) suggested—particularly because of the band at 463 mμ—the presence of triphenylene. This was confirmed by measuring the sample in IEPA (the broken curve). A considerable strengthening of the 0,0 band of the triphenylene at 430 mμ is observed. This effect is characteristic of triphenylene (compare the spectra of the pure hydrocarbon in EPA and IEPA given in Fig. 23).[2]

In considering the qualitative spectrophosphorimetric analysis of coal tar fractions, it should not be overlooked that the great majority of the compounds present in the tar has not so far been identified. The total number has been assessed at about 10,000. Of these some 5% have so far been recognized with certainty, and make up some 55% of the total weight.[3] In this situation it is understandable that many bands in the phosphorescence spectra of tar fractions cannot be assigned. It is quite conceivable that phosphorescence spectroscopy will occasionally lead to the discovery of new components in the tar.

[2] M. Zander, *Z. Anal. Chem.* **226**, 251 (1967).

[3] H. G. Franck, *Angew. Chem.* **63**, 260 (1951); *Brennstoff-Chem.* **45**, 5 (1964).

The quantitative spectrophosphorimetric determination of individual substances in tar fractions is carried out in the manner described elsewhere (see Section 3.4). Many of the compounds can be determined better by phosphorimetric methods than by UV or fluorescence spectroscopy. In Table 30 there are listed the spectroscopic methods by which

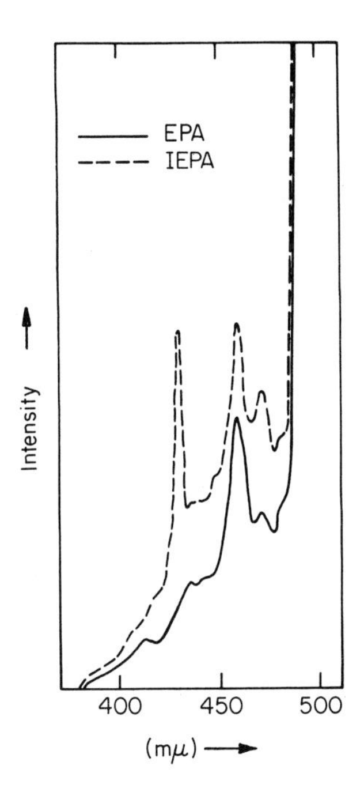

Fig. 33. Phosphorescence spectrum of a coal tar fraction on excitation with 300 mμ in EPA (—) and IEPA (--). [According to M. Zander, *Z. Anal. Chem.* **226**, 251 (1967).]

some of the constituents are best estimated quantitatively. The table also includes suitable excitation wavelengths and key bands for their phosphorimetric estimation.

The accuracy of the phosphorimetric evaluation of individual compounds in tar fractions may be revealed by the examination of test mixtures as well as by comparison with UV spectroscopic analyses. As an example there are given in Table 31 test analyses for the estimation of phenanthridone in tar fractions. The test mixtures were prepared by

TABLE 30 Spectroscopic Methods for the Determination of Individual Compounds in Coal Tar Fractions[a]

Compound	Method	Excitation wavelength for phosphorimetry (mμ)	Key band for phosphorimetry (mμ)
Anthracene	UV, fluorimetry		
Phenanthrene	Phosphorimetry	290	500
Carbazole	Phosphorimetry	290	408
Diphenylene sulfide	Phosphorimetry	320	425
Fluoranthene	Phosphorimetry	360	540
Pyrene	UV		
Phenanthridone	Phosphorimetry	325, 340	425
Triphenylene	Phosphorimetry	258	463
1,2-Benzanthracene	UV		
Chrysene	UV		
Tetracene	UV		
Peri-(1,8,9)-naphthoxanthene	Phosphorimetry, UV	398	515
3,4-Benzcarbazole	Phosphorimetry	360	483
Perylene	UV		
3,4-Benzpyrene	Fluorimetry		
1,2-Benzpyrene	Phosphorimetry	330	540

[a] M. Zander, *Erdoel Kohle* **19**, 278 (1966).

TABLE 31 Testing the Determination of Phenanthridone in Coal Tar Fractions[a]

Given (wt. %)	Found (wt. %)
2.0	2.1
3.0	2.9
4.0	3.9
5.0	5.0
7.0	6.7
9.0	9.1
12.0	11.6

[a] M. Zander, *Erdoel Kohle* **19**, 278 (1966).

adding phenanthridone in known quantities to tar fractions that contained none.

There are not many instances in which a compound in a tar fraction can be quantitatively evaluated with roughly the same accuracy both by phosphorimetry and by UV spectroscopy. One example is *peri*-(1,8,9) naphthoxanthene, which can be determined by UV spectroscopy using the band at 398 mμ and by means of its phosphorescence spectrum using that at 515 mμ. Ultraviolet and phosphorescence spectroscopic analyses for this compound were carried out for a series of tar fractions, and the results are quoted in Table 32.

TABLE 32 Comparison of Determinations Carried Out by UV and Phosphorescence Spectroscopy[a]

Sample number	*peri*-(1,8,9)-Naphthoxanthene (wt. %)	
	Phosphorescence	UV
24	2.6	2.3
26	4.1	4.2
27	5.2	4.9
28	6.9	6.8
30	10.4	9.4
33	8.6	10.3

[a] M. Zander, *Erdoel Kohle* **19**, 278 (1966).

4.3. Applications to the Study of Air Pollution

In the dust of the atmosphere (especially that of industrial towns), numerous polycyclic aromatic hydrocarbons and heterocyclics are to be found. It has been known for a long time that many of these compounds produce cancers in animals, and medical statistics have produced indirect evidence that the same is true also with human beings. It is therefore obvious why nowadays very great interest is shown in the analysis of airborne particles for the carcinogenic compounds present, to which all mankind is exposed. On this subject the work of E. Sawicki and his collaborators at the Robert A. Taft Engineering Center in Cincinnati, Ohio, is particularly important.

The analysis of samples of dust from the air is usually so carried out that the part that is soluble in an organic solvent is separated into coarse fractions either by column chromatography or by some other separation technique; each layer is then further resolved by paper or thin-layer chromatography. The separation proceeds either to the pure components or as far as mixtures that still consist of several substances. The pure components or mixtures so obtained must finally be qualitatively and quantitatively analyzed. For this purpose it is preferable to make use of spectroscopic methods, particularly fluorescence spectroscopy.

Sawicki[1] has pointed out, in a review that appeared in 1964, that phosphorescence spectroscopy should prove of inestimable service for trace analysis, and hence for the investigation of airborne dust samples. It is certain that the possible applications of spectrophosphorimetry in this field are far from exhausted. It follows, for example, that, for the recognition of aromatic and heterocyclic carbonyl compounds in samples of atmospheric dust, phosphorescence spectroscopy should be appreciably more effective than the better established methods.

Sawicki and Johnson[2] have shown that polycyclic aromatics from the air can easily be rendered visible on thin-layer chromatograms by means of their phosphorescence (in liquid nitrogen).

In the course of their studies Sawicki and his colleagues[3] compared the possibilities of applying UV, phosphorescence, and fluorescence spectroscopy to the characterization of polynuclear azaheterocyclic compounds. They found that, with the help of their phosphorescence spectra, the following isomers could be distinguished: phenanthridine and acridine, indeno(1,2,3-*ij*)isoquinoline and acenaphtho-(1,2-*b*)pyridine, benzo(*a*)acridine and benzo(*c*)acridine, dibenz(*a,h*)acridine and dibenz(*a, j*)acridine, and 14-phenyldibenz(*a, j*)acridine and 7-phenyldibenz(*c,h*)acridine. Spectrophosphorimetry showed, with the azaheterocyclics, considerably greater selectivity than UV spectroscopy. The reasons for this lie in the quite large differences in the phosphorescence quantum yields, the phosphorescence spectra, and phosphorescent lifetimes of the various compounds. As a disadvantage of the

[1] E. Sawicki, *Chemist-Analyst* **53**, 88 (1964).

[2] E. Sawicki and H. Johnson, *Microchem. J.* **8**, 85 (1964).

[3] E. Sawicki, T. W. Stanley, J. D. Pfaff, and W. C. Elbert, *Anal. Chim. Acta* **31**, 359 (1964).

phosphorescence technique it was asserted that the intensity measurements were difficult to reproduce. This, however, may have been attributable to the apparatus, since, in general, phosphorescence intensity measurements can be reproduced quite accurately.

4.4. Applications to the Analysis of Petroleum Products

Several examples of the application of phosphorescence spectroscopy to the complex field of petroleum analysis have appeared in the literature. Mamedov[1] has identified a number of polycyclic aromatic hydrocarbons in the wax distillates from petroleum by this means, among them being

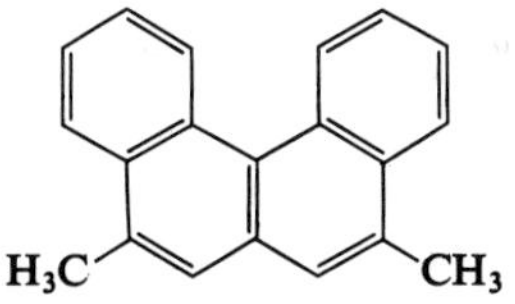

H_3C CH_3

CXV

2,9-dimethyl-3,4-benzphenanthrene (CXV). Khaluporskii[2] studied the phosphorescence and fluorescence of lubricating oils. He observed that products with good lubricating properties were recognizable by their low phosphorescences. Sidorov and Rodomakina[3] have reported on various applications of phosphorescence spectroscopy to the analysis of petroleum products. We shall now discuss in detail an important study of the application of spectrophosphorimetry to this problem by Drushel and Sommers.[4]

These authors investigated the azahydrocarbons and other nitrogen heterocyclics in petroleum products by combining the use of gas chromatography with UV, fluorescence, and phosphorescence spectroscopy. The basic nitrogen compounds from a straight-run middle distillate (430°–650°F) were extracted from the oil with dilute (1 : 1) hydrochloric acid, the extract was then neutralized with sodium hydroxide, and the

[1] Kh. J. Mamedov, *Izv. Akad. Nauk SSSR, Ser. Fiz.* **23**, 126 (1959); *Chem. Abstr.* **53**, 13,561g (1959).

[2] M. D. Khaluporskii, *Zavodsk. Lab.* **28**, 206 (1962); *Chem. Abstr.* **57**, 2494f and 8799d (1962).

[3] N. K. Sidorov and G. M. Rodomakina, *Uch. Zap. Saratovsk. Gos. Univ.* **69**, 161 (1960); *Chem. Abstr.* **57**, 12,782e (1962).

[4] H. V. Drushel and A. L. Sommers, *Anal. Chem.* **38**, 10 and 19 (1966).

nitrogen compounds that were precipitated in this way were dissolved in hexane. After the hexane solution was thoroughly washed with water, the solvent was distilled off and the basic nitrogen compounds remained behind. This material was used for the analytical examination. Gas chromatography effected a partial separation. The individual fractions leaving the gas chromatograph were collected in cooling traps and characterized spectroscopically. In the first fractions, quinolines seemed to be indicated by their UV spectra, but these overlap the spectra of substituted pyridines and tetrahydroquinolines, which could not therefore be confirmed. Phosphorescence spectra supplied additional evidence on the composition of the gas chromatographic fractions. By suitable selection of excitation wavelengths, it was possible to demonstrate the occurrence of quinolines and substituted pyridines or tetrahydroquinolines (or both) side by side in the fractions. Characterization was effected both by means of the spectra and by measurement of the phosphorescent lifetimes. Although quantitative analysis was not the object of the investigation, an estimate of the relative amounts of quinolines and substituted pyridines or tetrahydroquinolines or both in the specimens could be obtained. It showed that, as expected, pyridines or tetrahydroquinolines or both were concentrated in the first gas chromatographic fractions, and quinolines were concentrated in the later fractions. In these fractions acridine and benzacridines could also be detected through their intense fluorescence spectra.

A still more thorough study of this kind was carried out by Drushel and Sommers[4] on a light catalytic cycle oil with boiling range 400°–620°F. Starting from 3 kg of the oil and following a complicated process of working up in which, among other techniques, silica gel chromatography and extraction processes were employed, fractions were obtained in which carboxylic acids, phenols, basic nitrogen compounds, indoles, and carbazoles were concentrated. These fractions were examined—as described above—by gas chromatography and spectroscopy; the same gas chromatographic fractions were cleverly used for the measurement of infrared, phosphorescence, fluorescence, UV, and mass spectra.

The phosphorescence spectra gave more specific information than the fluorescence spectra. In the appropriate gas chromatographic fractions, substituted pyridines and quinolines could be detected via their phosphorescence spectra. Several spectra, possibly arising from tricyclic systems, remained unidentified. The indole–carbazole fraction produced two well-separated groups of peaks that were traced to substituted indoles and carbazoles. In the corresponding gas chromatographic fractions, compounds of the indole type were definitely detected. The carbazoles were also recognizable by their characteristic phosphorescence spectra. In every case the excitation spectra and the phosphorescence lifetimes were employed in the identification as well as the luminescence spectra.

TABLE 33 Investigation of the Nitrogen Compounds Resulting from the Catalytic Hydrogenation of Quinoline[a]

Gas chromatographic peak	Compound identified	Percentage by weight
1	Aniline	12.3
2	*o*-Toluidine	5.3
3	*o*-Ethylaniline	1.2
4	*N*-Ethyl-*o*-toluidine	0.3
5	*o*-Propylaniline (and other compounds)	4.2
6	Quinoline	47.8
7	2-Methylquinoline and 1,2,3,4-tetrahydroquinoline	14.1
8	3-Methylquinoline	1.3
9	Indole	4.4
10	2-Ethylquinoline	4.4
11	2-Isopropylquinoline	2.6
12	Dimers and codimers of partially hydrogenated quinolines, alkylquinolines and indoles	2.1
		100.0

[a] According to H. V. Drushel and A. L. Sommers, *Anal. Chem.* **38**, 10 (1966).

The combined application of gas chromatography and spectroscopy also justified itself in the investigation of the complicated mixture of reaction products obtained by the catalytic hydrogenation of quinoline. Here too, the great value of phosphorescence spectroscopy was clear, as had been stressed by Drushel and Sommers. By combination of the evidence provided by gas chromatography and by both phosphorescence and infrared spectroscopy on the individual gas chromatographic fractions, the majority of the compounds present were identified. The results of this interesting analysis are given in Table 33. It is worth comment that besides the normal hydrogenation of the quinoline system cracking and alkylation (with alkyl radicals from the cracking process) also take place. Attention may be called to the relatively high indole content of the mixture resulting from the catalysis.

4.5. Determination of Inhibitors in Polymers[1]

Inhibitors are often added to polymers to prevent oxidation and degradation, both of which alter the physical properties of the polymers in undesirable ways. The quantitative estimation of the inhibitors is an important analytical problem in the chemistry of synthetic products and so it is not surprising that much work has been published on it.

Frequently the inhibitor is extracted from the polymer before the real analysis is carried out. The determination of the inhibitor in the extract then takes place by IR or UV spectroscopy. However, the extraction process is usually tedious, frequently incomplete, and in many cases may change the inhibitor chemically as it proceeds. Consequently methods of evaluation are sought in which no extraction is necessary.

Such procedures are known. In them either IR or UV spectroscopy is used, but direct IR analysis of inhibitors in polymers has the disadvantage that the method is not very sensitive, and UV spectroscopic determination becomes difficult or impossible if the inhibitor displays no sharp, characteristic UV bands or if the absorption of the inhibitor is hidden beneath bands arising from the main component or additives. Hence use has been made in this field of the advantages that the luminescence spectroscopic methods possess compared with the absorption methods. In many instances the direct quantitative determination of inhibitors is possible by measuring the fluorescence spectra. Some inhibitors, however, show weak uncharacteristic fluorescences. As in other cases, weakly fluorescing inhibitors are frequently found to be intensely phosphorescent. Drushel and Sommers[1] have shown how they can be estimated.

For their experiments they used an ethylene–propylene rubber (EPR) and the inhibitors they studied were 2,2′-dimethyl-5,5′-di-*t*-butyl-4,4′-

[1] H. V. Drushel and A. L. Sommers, *Anal. Chem.* **36**, 836 (1964).

dihydroxydiphenyl sulfide (Santonox) and phenyl-2-naphthylamine (PBN). Santonox has no sharp absorption bands in the UV and only a very weak fluorescence. Its phosphorescence, on the other hand, is very strong. PBN can actually be quite well determined by UV spectroscopy, but difficulties arise from the overlapping of spectra if other additives are present in the synthetic material. PBN also shows intense phosphorescence. Thus a spectrophosphorimetric estimation procedure is obviously suggested for both these compounds.

In the procedure described by Drushel and Sommers, the phosphorescence of a thin film of the plastic that contains the inhibitor is measured at 77°K. The measurement was carried out with an Aminco-Keirs spectrophosphorimeter to which had been fitted a device for investigating thin luminescent films. The film thickness must be known; it can be found with a micrometer provided the film is not too thin. Because their extinctions of the transmitted light are too great, thicker films do not give linear phosphorescence calibration curves. Also inner filter effects caused by other components present in the polymer become more important with thicker films. Handling the thin films and the need to determine their thicknesses accurately are difficulties in this and similar methods.

In the paper by Drushel and Sommers, the phosphorescence spectra of the inhibitors investigated and the phosphorescence calibration curves for various film thicknesses are given. The accuracy of the method was studied during the determination of Santonox in EPR. Test analyses gave a relative error of 9%.

4.6. Applications in Biochemistry and Pharmacology

The first example ever given of the analytical application of phosphorescence spectra to organic compounds came from the field of biochemistry. In 1955 Rybak *et al.*[1] proposed phosphorescence spectroscopy as a method of analysis for amino acids. Their publication, however, contained few experimental details. Since then a series of communications has appeared from various authors concerning the applications of spectrophosphorimetry in biochemistry and pharmacology.

Freed and Vise[2] developed the ideas of Rybak, Lochet, and Rousset[1]

[1] B. Rybak, R. Lochet, and A. Rousset, *Compt. Rend.* **241**, 1278 (1955).

[2] S. Freed and H. M. Vise, *Anal. Biochem.* **5**, 338 (1963).

further and used spectrophosphorimetry for the determination of the proteolytic enzyme α-chymotrypsin. The phosphorescence of this substance arises from its tryptophan component (CXVI). A mixture of water, methyl alcohol, and ether in the proportions of 5:11:4 by volume was used as solvent, and the measurement was made at 113°K. The reproducibility of the determination was found to depend essentially on the purity of the solvent. This publication also gives the emission and excitation spectra of *N*-acetyl-L-tyrosine ethyl ester (CXVII), which can likewise be spectrophosphorimetrically determined.

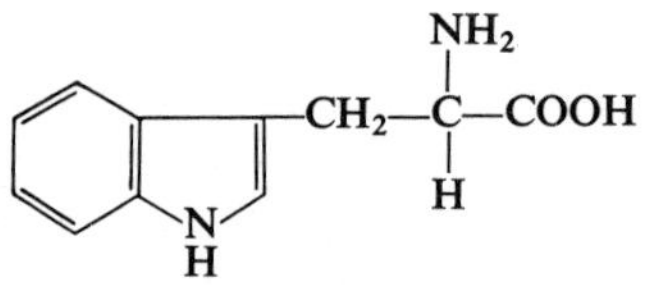

CXVI

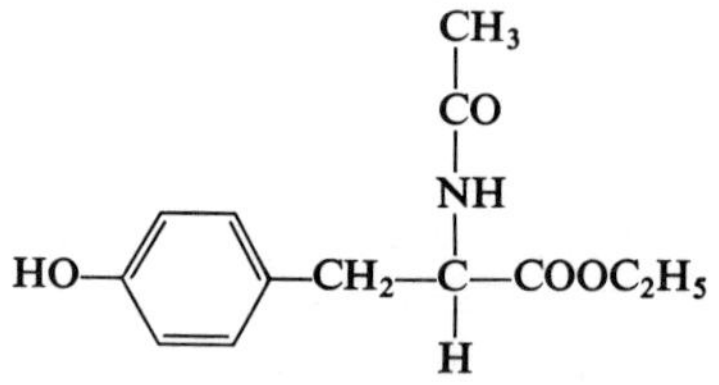

CXVII

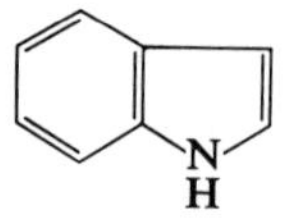

CXVIII

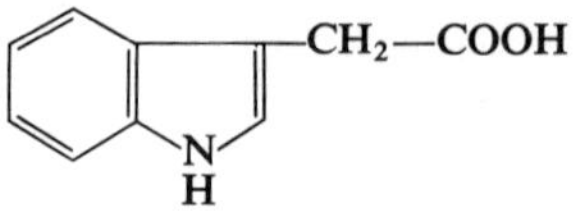

CXIX

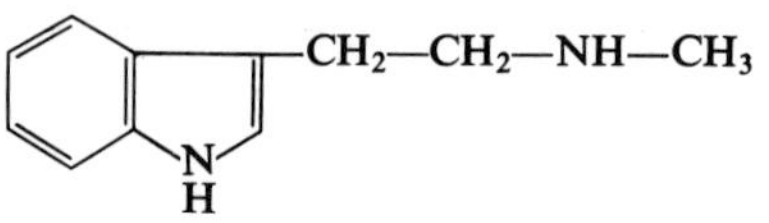

CXX

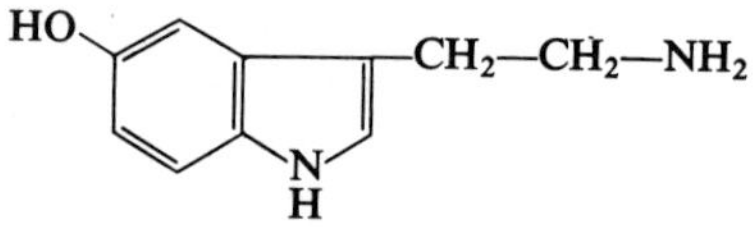

CXXI

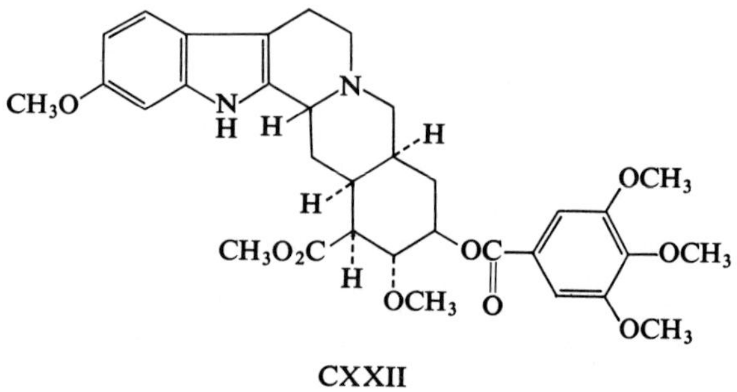

CXXII

Freed and Salmre[3] used phosphorescence spectroscopy to characterize several biochemically and pharmacologically important indole derivatives. The phosphorescence spectra of indole (CXVIII), indoleacetic acid (CXIX), tryptophan (CXVI), tryptamine (CXX), serotonin (CXXI), and reserpine (CXXII) were examined. The solvent was a methanol–ethanol mixture. The spectra of several of the substances studied could also be obtained in frozen aqueous solutions provided electrolytes were present. The most important result of this work is that the phosphorescence spectra of the indole derivatives are substantially more characteristic than their fluorescence spectra. The latter consist, at room temperature, of a single broad band the position of which depends only slightly on the structure of the compound. At low temperature there is a hint of vibrational structure, but this is nothing like so well developed as in the phosphorescence spectra, which, in addition, may vary somewhat from compound to compound. Thus although the derivatives investigated cannot be distinguished by their fluorescence spectra, it should be possible to identify them by means of their phosphorescence spectra. They can in fact also be detected about ten times as sensitively by phosphorescence as by fluorescence.

We owe some interesting applications of spectrophosphorimetry in biochemistry, pharmacology, and clinical chemistry to Winefordner and his colleagues.[4] These authors found that it is frequently possible to determine foreign substances such as drugs and alkaloids in blood serum, plasma, or urine rapidly, accurately, and simply by spectrophosphorimetry. Winefordner and Latz[5] showed this first of all with respect to the quantitative evaluation of aspirin (acetylsalicylic acid) in blood serum and plasma.

The determination of aspirin in blood and other biological material is important in clinical chemistry and is usually carried out colorimetrically by means of the ferric salicylate complex. For this, aspirin is first converted into salicylic acid by hydrolysis. The method is liable to error with low concentrations of aspirin, requires a high background correction if the quantity of salicylate is less than 10 mg/100 ml of serum, and is tedious to carry out. Winefordner's method, on the other hand, is specific for

[3] S. Freed and W. Salmre, *Science* **128**, 1341 (1958).

[4] J. D. Winefordner, *in* "Fluorescence and Phosphorescence Analysis" (D. M. Hercules, ed.), p. 176ff. Wiley (Interscience), New York, 1966.

[5] J. D. Winefordner and H. W. Latz, *Anal. Chem.* **35**, 1517 (1963).

aspirin and both rapid and accurate over a wide range of concentrations.

The applicability of a spectrophosphorimetric method of estimating aspirin in blood naturally presupposes that the serum and plasma and the substances normally present in them do not phosphoresce at all or do so much more weakly than the aspirin. This question has been studied very thoroughly by Winefordner and Latz. Of the substances investigated, thiamine, riboflavin, tyrosine, and tryptophan showed measurable phosphorescence, but cholesterol, bilirubin, thyroxine, uric acid, creatinine, glucose, urea, and vitamin C did not. The amino acid tryptophan gave the most intense phosphorescence. The intensity of a saturated solution of this compound in EPA is comparable with that of a solution which contains 0.2 mg of aspirin in 100 ml of EPA. The EPA extracts of serum gave broad phosphorescence emissions corresponding closely with that of tryptophan, and the measurements could be reproduced quite well. Chloroform extracts likewise gave only the tryptophan phosphorescence. When Winefordner's method for determining aspirin was applied to aspirin-free samples of serum, the observed phosphorescence corresponded to an apparent aspirin concentration of 0.5 mg/100 ml of serum. It follows, therefore, from these investigations that the phosphorescence of the aspirin is considerably more intense than that of any other natural component of blood. Therefore, only when estimating relatively low concentrations of aspirin is it necessary to apply any background correction.

The aspirin in blood is hydrolyzed to salicyclic acid at an appreciable rate and hence the influence of salicyclic acid on the aspirin determination must be known. It has been shown that salicyclic acid has a phosphorescence which is only about one five-hundredth as strong as that of aspirin. It can, therefore, cause no significant interference with the evaluation of aspirin. Winefordner and Latz used for their measurements a spectrophosphorimeter they had built themselves. Their method can, however, be carried over to other similar instruments, such as the Aminco-Keirs spectrophosphorimeter. The white, unfiltered light of a xenon lamp was used for the excitation. Aspirin has a single, broad phosphorescence band with its maximum at 410 mμ. All phosphorescence measurements are carried out with this wavelength. First a phosphorescence calibration curve is set up to give the intensity of phosphorescence versus concentration of aspirin (in EPA). It covers a range of concentration from 0.01 to 100 mg of aspirin in 100 ml of EPA, and is linear over

2 powers of 10, deviating from linearity at higher concentrations only slightly (see Fig. 34). For the determination of aspirin in blood serum or plasma, 0.4-ml samples are taken in 10-ml glass-stoppered graduated cylinders to which 0.1 ml of concentrated hydrochloric acid and 7.5 ml of chloroform are added. The mixture is shaken vigorously for 30 seconds

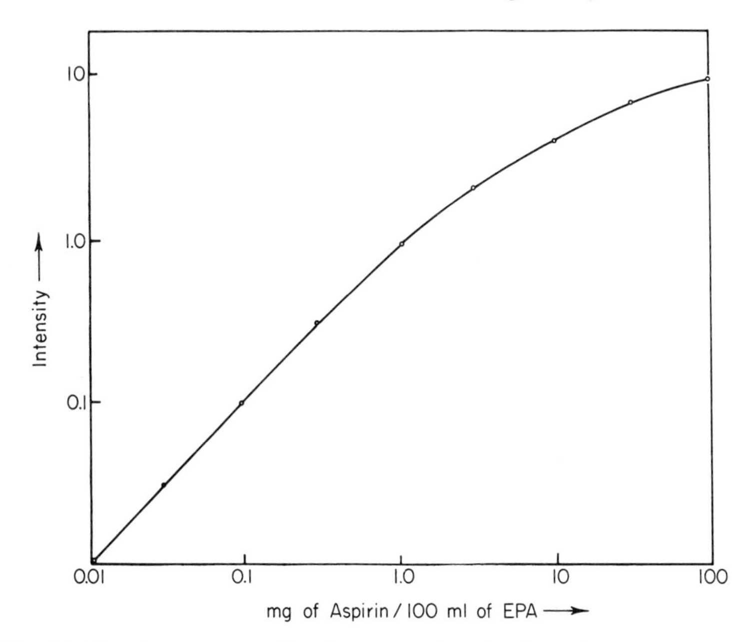

Fig. 34. Phosphorescence calibration curve for the determination of aspirin [according to J. D. Winefordner and H. W. Latz, *Anal. Chem.* **35**, 1517 (1963)].

and then the chloroform layer is separated; 1.00 ml of this is transferred to a 10 ml beaker and the chloroform is driven off in a stream of air. The residue is taken up in 1.00 ml of EPA and the resulting solution is used for the phosphorescence measurement as described above. From the intensity value thus obtained there may be subtracted, if necessary, the phosphorescence blank value of the EPA used and that of an aspirin-free sample of serum or plasma. From the corrected value the concentration of aspirin in the EPA is obtained with the help of the calibration curve and is then multiplied by 18.75 to obtain the concentration of aspirin (in milligrams) in 100 ml of plasma or serum.

The accuracy of the method has been established by test analyses. For these, known quantities of aspirin were weighed into aspirin-free serum or plasma and the mixtures were analyzed in the way described above. The results of these test analyses have been collected in Table 34. They

TABLE 34 Test Analyses for the Determination of Aspirin in Blood Plasma and Serum[a]

a. In Citrated Plasma

Aspirin in 100 ml plasma (mg)		Relative standard deviation (%)	Recovery (%)
Given	Found (mean of 4 or 5 analyses)		
1.00[b]	0.95	8	95
5.00[b]	5.00	13	100
10.00[b]	10.10	8	101
25.00[b]	23.00	5	92
50.00[b]	43.00	10	86

b. In Pooled Serum

Aspirin in 100 ml serum (mg)		Recovery (%)
Given	Found (mean of 3 analyses)	
1.0	1.1	110
1.5[b]	2.0	133
5.0	4.2	84
7.5	7.4	99
10.0	10.3	103
15.0[b]	14.0	93
25.0	26.0	104
37.0[b]	28.0	76
50.0	41.0	82

[a] J. D. Winefordner and H. W. Latz, *Anal. Chem.* **35**, 1517 (1963).
[b] These samples also contained an equal amount of salicylic acid.

show a relative standard deviation of 5–10%. The limits of detection for the method lie at about 0.2 mg of aspirin per 100 ml of serum or plasma. The time needed for a complete analysis amounted to less than 10 minutes.

As Winefordner and Tin[6] have shown, spectrophosphorimetry can also be brought into use for the quantitative determination of numerous other medically important substances in biological media. These authors have described in detail methods for the estimation of procaine (CXXIII), phenobarbital (CXXIV), cocaine (CXXV), and chlorpromazine in blood serum as well as of cocaine and atropine (CXXVI) in urine. In each case the substance is extracted from the biological fluid with chloroform or ether, the solvent is evaporated, and the residue dissolved in EPA, and the resulting solution measured phosphorimetrically.

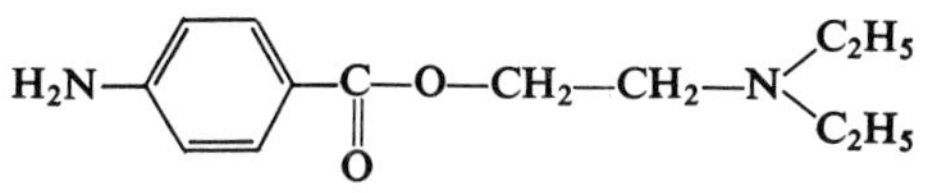

$H_2N-C_6H_4-C(=O)-O-CH_2-CH_2-N(C_2H_5)_2$

CXXIII

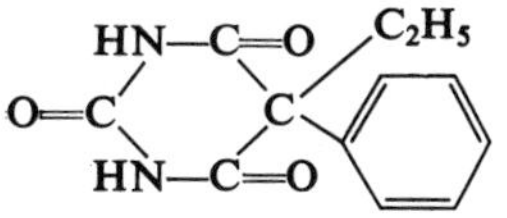

(HN—C=O)(HN—C=O) O=C ... C(C_2H_5)(C_6H_5)

CXXIV

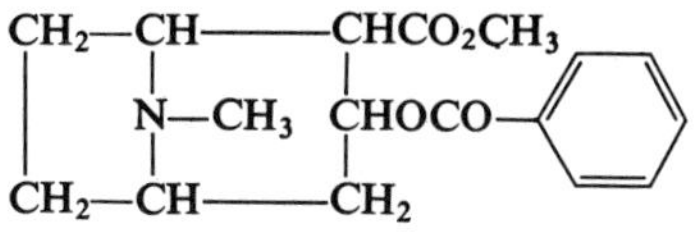

CH₂—CH——CHCO₂CH₃ / N—CH₃ CHOCO—C₆H₅ / CH₂—CH——CH₂

CXXV

CH₂—CH——CH₂ CH₂OH / N—CH₃ CHOCOCH—C₆H₅ / CH₂—CH——CH₂

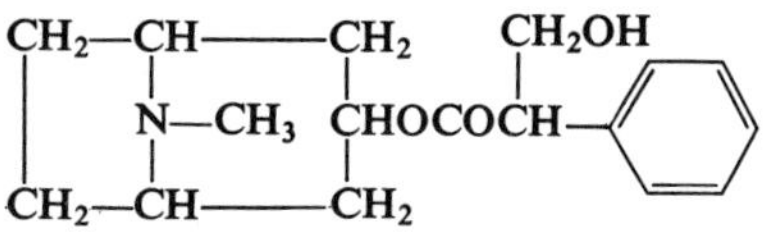

CXXVI

As in the case of aspirin, the phosphorescences of the pharmaceuticals investigated by Winefordner and Tin[6] are substantially more intense than the background phosphorescence of the blood serum or plasma. The same is true with urine if the extraction is performed with ether and the pH value is kept greater than 5. Possible interference from the presence of metabolic products of the drugs being investigated was thoroughly discussed at the same time. The only metabolite of procaine that shows phosphorescence is *p*-aminobenzoic acid. Addition of small quantities of *p*-aminobenzoic acid to the procaine caused no significant change in the procaine analysis. The reason is obviously that the *p*-aminobenzoic acid has a very small distribution coefficient between

[6] J. D. Winefordner and M. Tin, *Anal. Chim. Acta* **32**, 64 (1965); see also H. A. Moye and J. D. Winefordner, *J. Agr. Food Chem.* **13**, 533 (1965).

chloroform and water at the pH value of 10 used in the analysis. Under these conditions, therefore, *p*-aminobenzoic acid is not extracted at all from the blood serum by the chloroform and so does not appear in the solution for analysis. The behavior of the possible metabolic products of the other drugs examined is very similar.

The analytical procedures described by Winefordner and Tin refer solely to the determination, in each case, of one particular drug in the blood or urine. In practice several drugs will frequently be present. In many such cases the desired objective will be achieved by applying the phosphorimetric techniques for the analysis of many-component systems.

In Table 35 there are collected the phosphorescence excitation wavelengths, key bands, limits of detection, and relative standard deviations for the determination of procaine, cocaine, phenobarbital, atropine, and chlorpromazine in urine or blood.

In a further contribution Winefordner and Tin[7] have published

TABLE 35 Spectrophosphorimetric Determination of a Number of Drugs in Urine and Blood[a]

Compound	Medium	Excitation wavelength (mμ)	Key band (mμ)	Detection limit[b] (gm/ml)	Relative standard deviation[c] (%)
Procaine (CXXIII)	Blood	310	430	1×10^{-8}	2–5
Cocaine (CXXV)	Blood	240	400	1×10^{-8}	3–10
Cocaine	Urine	240	400	1×10^{-8}	2–10
Phenobarbital (CXXIV)	Blood	240	380	1×10^{-7}	2–3
Atropine (CXXVI)	Urine	240	380	1×10^{-7}	2–3
Chlorpromazine hydrochloride	Blood	320	490	—	3–4

[a] J. D. Winefordner and M. Tin, *Anal. Chim. Acta* **31**, 239 (1964); **32**, 64 (1965).
[b] The detection limits quoted refer to solutions in ethanol.
[c] To determine the relative standard deviations the test samples were each investigated five times. The standard deviation varies with the range of concentrations, so the figures given above are limiting values.

[7] J. D. Winefordner and M. Tin, *Anal. Chim. Acta* **31**, 239 (1964); see also H. C. Hollifield and J. D. Winefordner, *Talanta* **14**, 103 (1967).

phosphorescence data for numerous medically important compounds without specifying the methods of estimating them in biological media. The possibility of analyses of this kind has, however, been discussed and the appropriate techniques follow by analogy with those described in the other publications of Winefordner. Alcohol was used as the solvent in these measurements because the solubility of many drugs in alcohol is considerably greater than in EPA. Table 36 gives the phosphorescence excitation wavelengths, phosphorescence bands, lifetimes, and limits of detection for some of the compounds examined by Winefordner and Tin.[7]

TABLE 36 Spectrophosphorimetric Determination of Various Drugs[a]

Compound[b]	Excitation wavelength (mμ)	Bands[c] (mμ)	Lifetime (sec)	Detection limit (gm/ml)
Mebaral	240	380	2.2	1×10^{-8}
Rutonal	240	380	2.5	2×10^{-8}
Benzocaine	310	430, 420, 440	5.3	7×10^{-9}
p-Aminobenzoic acid	310	430, 420, 440	3.2	4×10^{-9}
Butacaine sulfate	310	430, 420, 440	5.7	5×10^{-8}
Cyclaine hydrochloride	240, 290	400, 410, 370	2.4	6×10^{-9}
Metycaine hydrochloride	240, 290	400, 410, 370	2.7	6×10^{-9}
Benzoic acid	240, 290	400, 410, 370	2.3	5×10^{-9}
Quinidine sulfate	340, 250	500, 470	1.3	5×10^{-8}
Quinine hydrochloride	340, 250	500, 470	1.3	4×10^{-8}
Lidacaine	265, 240	400	1.1	1.2×10^{-6}
Caffeine	285, 245	440	2.0	2×10^{-7}
Ephedrine	225, 410	390	3.6	2×10^{-7}
Phenylephrine hydrochloride	290, 240	390	2.4	1×10^{-8}
Tronothane hydrochloride	300, 240	410	1.2	2×10^{-8}
Cinchophen	350, 270	520, 490	0.8	2×10^{-8}
Physostigmine sulfate	315, 260	420	3.6	3×10^{-8}
Chlortetracycline	280	410	2.7	5×10^{-8}

[a] J. D. Winefordner and M. Tin, *Anal. Chim. Acta* **31**, 239 (1964).

[b] The compounds are arranged according to their structural similarities and the relationships between their spectral characteristics.

[c] The first band given is suitable for use as key band, and the phosphorescent lifetime and limits of detection have been determined for this.

TABLE 37 Spectrophosphorimetric Determination of Various Alkaloids[a, b]

Compound	Excitation wavelength (mμ)	Bands (mμ)	Lifetime (sec)	Detection limit (gm/ml)
Codeine	275	505	0.3	1×10^{-8}
Morphine	285	500	0.25	1×10^{-8}
Papaverine hydrochloride	245, *305*, 360	455, *480*	1.5	5×10^{-10}
Yohimbine hydrochloride	290	410	7.4	1×10^{-8}
Apomorphine hydrochloride	320	440, *470*	3.1	1×10^{-9}
Narceine	290	440	0.5	1×10^{-7}
Thebaine	315	500	1.0	1×10^{-6}
Brucine	305	435	0.88	1×10^{-7}
Strychnine phosphate	220, *290*	325, *440*	1.2	5×10^{-5}
Morphine sulfate	265	460	0.75	1×10^{-5}
Narcotine	315	440	0.5	1×10^{-8}

[a] H. C. Hollifield and J. D. Winefordner, *Talanta* **12**, 860 (1965).
[b] The data for morphine sulfate and narcotine relate to acidified alcoholic solutions; all the rest, to neutral alcoholic solutions. Italics indicate the most intense band in each case.

An investigation by Hollifield and Winefordner[8] is particularly concerned with the phosphorescence of alkaloids. The quantitative determination of alkaloids in medicines, plants, and biological media is an analytical problem of medical and forensic importance. Here spectrophosphorimetry presents itself as a very useful method of investigation. The phosphorescence quantum yields of the alkaloids are relatively high and the limits of detection are correspondingly low. It was found that all the alkaloids examined could be detected more sensitively in neutral than in either acid or alkaline solution. A limitation of the method arises from the fact that both the excitation spectra and the phosphorescence spectra consist of only a few broad bands that easily become superposed in mixtures. In analyzing such mixtures phosphorimetrically, it will therefore frequently be necessary first to separate the individual components from each other by suitable methods such as thin-layer chromatography, and to employ phosphorimetry only for

[8] H. C. Hollifield and J. D. Winefordner, *Talanta* **12**, 860 (1965).

determining quantitatively the pure compounds isolated. However, one can conceive of mixtures of alkaloids that can be analyzed entirely by phosphorimetry through suitable choice of excitation wavelengths and key bands.

Excitation wavelengths, phosphorescence bands (in ethanol), phosphorescent lifetimes, and limits of detection (in grams per milliliter) are given in Table 37 for the alkaloids investigated by Hollifield and Winefordner.[8]

4.7. Applications in Food Chemistry and Related Fields

McCarthy and Winefordner[1] have reported an interesting application of spectrophosphorimetry to food chemistry: the rapid, quantitative determination of biphenyl in oranges.

Biphenyl is frequently used as a fungicide for citrus fruits to prevent perishing during storage and transportation and is found not only on the peel but also in the flesh. As in many countries limits of tolerance for biphenyl in citrus fruits are specified, a rapid and accurate method for its estimation is of great importance. Many procedures have been described in the literature. In all of them steam distillation is employed to isolate the fungicide from the fruit and UV and IR spectroscopy as well as colorimetric and gas chromatographic techniques are then used for the quantitative determination of the compound in the distillate. There are many shortcomings in all these procedures: the amounts of test material required are relatively large, for some processes up to 2 pounds for a single estimation; the time taken for the analyses is quite long because of the steam distillation, in some cases up to 48 hours; the accuracy of many of the procedures that are applied leaves much to be desired and they are not all specific. In contrast the method described by McCarthy and Winefordner[1] represents a very important advance.

For the isolation of biphenyl from the fruit, Winefordner's method substitutes extraction with ether for steam distillation. Normally the analyses can be carried out even on a single orange, although it is desirable to use several in order to obtain a better average. The oranges are first quartered and then separated into pulp and juice on the one hand and peel on the other. The material is homogenized and then the

[1] W. J. McCarthy and J. D. Winefordner, *J. Assoc. Offic. Agr. Chemists* **48**, 915 (1965).

pulp and juice and the separated peel undergo several extractions with ether according to instructions given in detail in Winefordner's paper.[1]

The ether extracts, which have taken up the biphenyl from the fruits quantitatively, are then subjected to thin-layer chromatography. For this, aluminum oxide G is used as the adsorbent. This material includes phosphorescent portions if it is used untreated; it is therefore heated to a high temperature for 2 hours before use. 2-Methylbutane is the approved solvent for the thin-layer chromatography and the running time of the chromatogram is about 15 minutes. The compounds isolated in this way are rendered visible by passing iodine vapor over them. Biphenyl appears as a dark orange-colored stripe at an R_f value of 0.6. The biphenyl spot from a second chromatogram that has not been treated with iodine is scraped off, taken up in alcohol, and the alcoholic solution is measured phosphorimetrically.

This spectrophosphorimetric determination procedure has been described for the Aminco-Keirs spectrophosphorimeter. The excitation wavelength 275 mμ is used and the key band is the biphenyl band at 470 mμ. First calibration curves of phosphorescence intensity versus concentration are set up in the usual way with solutions containing known concentrations of biphenyl in ethanol. The calibration curve extends over a range from 10^{-3} to 10^{2} μg/ml. The sample obtained from the thin-layer chromatography is then measured phosphorimetrically; the biphenyl concentration in the alcohol is ascertained and converted into the quantity present in the oranges.

As Winefordner's method for estimating biphenyl in oranges differs substantially and in many respects from the previously known methods, a series of fundamental questions connected with the procedure had to be cleared up in order to exclude definitely the possibility of systematic errors. First it was important to establish conclusively whether the ether extraction procedure, which saves a great deal of time, adequately replaces the steam distillation. In a series of test experiments known quantities of biphenyl were added to oranges that were known to be free from this substance, and the fruits so prepared were analyzed by the new method. It was found that 95–100% of the quantities of biphenyl added were recovered. In a second series of experiments, parallel investigations were carried out in which on the one hand the biphenyl was isolated from the oranges by steam distillation and on the other by ether extraction, the resulting distillates and extracts being then examined to estimate the biphenyl. The results agreed remarkably well. Also the question of whether other substances that are present in oranges interfere with the phosphorimetric determination of biphenyl has been thoroughly studied. When oranges which it was certain contained no biphenyl were investigated, no spots appeared in the thin-layer chromatogram in the position where the biphenyl should have been. Conversely the phosphorescence excitation and emission spectra of biphenyl that had been isolated from oranges by ether extraction and thin-layer chromatography were entirely identical with those of pure biphenyl. To sum up, it is clear that Winefordner's method is not subject to systematic errors.

The accuracy of the method is about ±0.2 ppm with 3.8–8.9 ppm of biphenyl in the juice and pulp and about ±2 ppm with 39–65 ppm of biphenyl in the peel. A complete analysis can be carried out (on four oranges) with several controls in about 2 hours. The application of the method is not restricted to oranges, but could easily be extended to other citrus fruits on which biphenyl has been used as a fungicide.

The determination of pesticides is, from the practical point of view, an important problem in analytical chemistry and also has significance, for example, in food chemistry and forensic medicine. That spectrophosphorimetry can be successfully introduced into this field too has been shown by Moye and Winefordner.[2] These authors examined the phosphorescence of 52 pesticides (including several decomposition products). Thirty-two of them showed measurable phosphorescence, and their excitation spectra, phosphorescence spectra, lifetimes, calibration curves, and limits of detection could be estimated. A summary of the results obtained is given in Table 38. No measurable phosphorescence was shown by the following: Chlordan, Endrin, Heptachlor, Lindane, Methylparathion, Malathion, Thimet, Thiodan, Delnav, HEOD, HHDN, Aldrin, Dieldrin, Telodrin, 3-amino-1,2,4-triazole, Phosphamidon, U.C.21149, Dimethoate, Dimethoate acid, and oxygen analogs of Dimethoate.

For several of the compounds included in Table 38 the limits of detection by phosphorescence are very low. They ought to be detected more sensitively by phosphorimetry than by other analytical methods. This is the case, for example, with Co-Ral and *p*-nitrophenol, of which even 50 pg/ml can be recognized phosphorimetrically. The limits of detection of Parathion ought to be 2 powers of 10 lower if it is converted

[2] H. A. Moye and J. D. Winefordner, *J. Agr. Food Chem.* **13**, 516 (1965).

into *p*-nitrophenol by hydrolysis before it is determined quantitatively. The same is true of methyl parathion, which does not phosphoresce measurably itself.

Another interesting investigation by Winefordner and Moye[3] was concerned with the quantitative estimation of the tobacco alkaloids nicotine (CXXVII), nornicotine (CXXVIII), and anabasine (CXXIX). For this analytical problem several procedures were already available, but they were either quite specific and accurate but tedious to carry out, or else could be rapidly carried out but then were less specific and accurate. In contrast, Winefordner's method shows considerable advantages over those known longer.

N CH_3

CXXVII

N H

CXXVIII

N H

CXXIX

TABLE 38 Spectrophosphorimetric Determination of Pesticides and Related Compounds[a, b]

Compound	Excitation wavelength (mμ)	Bands[c] (mμ)	Lifetime (sec.)	Detection limit (gm/ml)	Linear range of calibration curve (moles/liter)
DDT (*p,p′*)	270	420	0.2	7×10^{-10}	1.9×10^{-9}–7.4×10^{-4}
DDD (*p,p′*)	265	415	0.2	1×10^{-9}	4.0×10^{-9}–1.3×10^{-3}
DDE (*p,p′*)	270	425	0.2	2×10^{-10}	1.0×10^{-9}–8.8×10^{-4}
Kelthane	285	515	0.2	6×10^{-10}	1.5×10^{-9}–7.1×10^{-4}
Methoxychlor	275	380, 395, 360	0.7	4×10^{-10}	1.3×10^{-9}–9.6×10^{-5}
Chlorobenzilate	275	415, 425, 445, 400, 480	0.2	1×10^{-9}	3.0×10^{-9}–1.2×10^{-3}

[3] J. D. Winefordner and H. A. Moye, *Anal. Chim. Acta* **32**, 278 (1965).

TABLE 38—*continued*

Compound	Excitation wavelength (mμ)	Bands[c] (mμ)	Lifetime (sec.)	Detection limit (gm/ml)	Linear range of calibration curve (moles/liter)
Toxaphene	240	390	1.9	2×10^{-8}	4.5×10^{-8}–7.5×10^{-3}
Kepone	260	410	1.25	1×10^{-6}	2.0×10^{-6}–9.2×10^{-3}
Sulfenone	275	390, 375	0.2	5×10^{-10}	2.0×10^{-9}–9.1×10^{-5}
Tedion	295	410	0.2	2×10^{-10}	5.0×10^{-8}–6.3×10^{-5}
Orthotran	260	395, 375	<0.2	2×10^{-9}	8.0×10^{-9}–7.5×10^{-4}
Parathion	360	515, 490	<0.2	8×10^{-9}	3.0×10^{-8}–9.0×10^{-4}
Ronnel	300	475	<0.2	6×10^{-10}	2.0×10^{-9}–1.0×10^{-3}
Co-Ral	335	510, 490	<0.2	4×10^{-11}	1.0×10^{-10}–8.3×10^{-6}
Diazinon	275	395, 375	5.0	3×10^{-9}	1.0×10^{-8}–1.1×10^{-3}
Guthion	325	420, 400	0.6	6×10^{-8}	2.0×10^{-7}–7.5×10^{-3}
Trithion	305	430	<0.2	3×10^{-10}	8.0×10^{-10}–8.5×10^{-5}
Aramite	285	400	3.3	3×10^{-10}	1.0×10^{-9}–1.1×10^{-4}
Isolan	285	395	1.6	2×10^{-7}	1.0×10^{-6}–1.3×10^{-2}
Sevin	300	510, 475, 485, 550	2.0	4×10^{-9}	2.0×10^{-8}–1.0×10^{-3}
Zectran	285	440	0.45	5×10^{-9}	2.5×10^{-8}–7.6×10^{-4}
Bayer 44646	290	460	0.60	1×10^{-8}	6.0×10^{-8}–6.3×10^{-4}
Bayer 37344	275	435	<0.2	1×10^{-8}	5.0×10^{-8}–7.4×10^{-3}
NIA 10242	285	400	1.6	7×10^{-10}	3.0×10^{-9}–7.3×10^{-4}
U.C. 10854	270	385	2.9	2×10^{-9}	1.0×10^{-8}–1.2×10^{-3}
Imidan	305	440, 420	0.75	6×10^{-10}	2.0×10^{-9}–8.5×10^{-5}
2,4,5-Trichlorophenoxyacetic acid	300	480	<0.2	5×10^{-10}	2.0×10^{-9}–9.5×10^{-4}
2,4-Dichlorophenoxyacetic acid	290	495	<0.2	4×10^{-9}	2.0×10^{-8}–8.6×10^{-4}
p-Chlorophenol	290	505	<0.2	2×10^{-8}	1.8×10^{-7}–1.1×10^{-2}
2,4,5-Trichlorophenol	305	485	<0.2	3×10^{-9}	1.5×10^{-8}–6.7×10^{-4}
p-Nitrophenol	355	520, 495	<0.2	2×10^{-11}	1.7×10^{-10}–5.0×10^{-4}
1-Naphthol	320	475, 495, 520	1.15	2×10^{-10}	1.7×10^{-9}–8.4×10^{-4}

[a] H. A. Moye and J. D. Winefordner, *J. Agr. Food Chem.* **13**, 516 (1965).
[b] All measurements were made in absolute alcohol.
[c] The bands quoted first are suitable for use as key bands.

In Winefordner's method the tobacco sample is first finely powdered and then extracted with chloroform under alkaline conditions. The chloroform extract, which has taken up the tobacco alkaloids quantitatively, is then submitted to thin-layer chromatography. For this aluminum oxide G is employed as the adsorbent and a mixture of 100 ml of chloroform and 1.5 ml of methanol as the solvent system. The running time for the chromatogram is about 30 minutes. Several thin-layer chromatograms are allowed to run in parallel on the same sample of tobacco and are dried at the end of the development. On one of the chromatograms the spots of alkaloid are rendered visible by spraying with Dragendorff's reagent, when they become recognizable by their dark orange color on the yellow background. On a second chromatogram, which has not been treated with Dragendorff's reagent, the compounds are located by means of their R_f values. The corresponding regions are scraped from the plate and dissolved in a mixture of alcohol and sulfuric acid. An aliquot part of this solution is employed for the spectrophosphorimetric examination.

Since the absorption and phosphorescence spectra of the three tobacco alkaloids coincide very extensively, direct spectrophosphorimetric analysis of a mixture of the compounds is not possible and the chromatographic separation described above is therefore necessary. The excitation wavelength for all three alkaloids is 270 mμ and the key band that at 390 mμ. Phosphorescence calibration curves are prepared with the pure compounds under the same conditions. The percentage of an alkaloid in the tobacco sample is then calculated from the formula

$$\%\ \text{Alkaloid} = \frac{CM_W FV}{W} \times 100\% \tag{15}$$

where C is the concentration of the alkaloid in moles per liter of the alcohol–sulfuric acid solution as given by the phosphorimetry, M_W the molecular weight of the alkaloid in question, F the factor allowing for the dilution of the sample (1000 if 100 ml of chloroform extract was prepared and only 0.1 ml used for the chromatographic separation), V the volume, in liters, used to remove the alkaloid from the thin-layer (usually 0.005 liter), and W the weight of the tobacco sample.

Test analyses gave a maximum standard deviation of 6%. The total time to estimate the three alkaloids in a sample of tobacco came to less than 90 minutes.

AUTHOR INDEX

Numbers in parentheses are reference numbers and indicate that an author's work is referred to, although his name is not cited in the text.

E

F

G

H

R

S

T

U

V

W

Y

Z

SUBJECT INDEX

Q

R